Translational Systems Sciences

Volume 2

More information about this series at http://www.springer.com/series/11213

In 1956, Kenneth Boulding explained the concept of General Systems Theory as a *skeleton of science*. The hope was to develop something like a "spectrum" of theories—a system of systems which might perform the function of a "gestalt" in theoretical construction. Such "gestalts" in special fi elds have been of great value in directing research towards the gaps which they reveal.

There were, at that time, other important conceptual frameworks and theories, including cybernetics. Additional theories and applications developed later, such as synergetics, cognitive science, complex adaptive systems, and many others. Some focused on principles within specifi c domains of knowledge and others crossed areas of knowledge and practice, along the spectrum described by Boulding.

Also in 1956, the Society for General Systems Research (now the International Society for the Systems Sciences) was founded. One of the concerns of the founders, even then, was the state of the human condition, and what science could do about it.

The present Translational Systems Sciences book series aims at cultivating a new frontier of systems sciences for contributing to the need for practical applications that benefit people.

The concept of translational research originally comes from medical science for enhancing human health and well-being. Translational medical research is often labeled as "Bench to Bedside." It places emphasis on translating the findings in basic research (*at bench*) more quickly and efficiently into medical practice (*at bedside*). At the same time, needs and demands from practice drive the development of new and innovative ideas and concepts. In this tightly coupled process it is essential to remove barriers to multi-disciplinary collaboration.

The present series attempts to bridge and integrate basic research founded in systems concepts, logic, theories and models with systems practices and methodologies, into a process of systems research. Since both bench and bedside involve diverse stakeholder groups, including researchers, practitioners and users, translational systems science works to create common platforms for language to activate the "bench to bedside" cycle.

In order to create a resilient and sustainable society in the twenty-first century, we unquestionably need open social innovation through which we create new social values, and realize them in society by connecting diverse ideas and developing new solutions. We assume three types of social values, namely: (1) values relevant to social infrastructure such as safety, security, and amenity; (2) values created by innovation in business, economics, and management practices; and, (3) values necessary for community sustainability brought about by confl ict resolution and consensus building.

The series will first approach these social values from a systems science perspective by drawing on a range of disciplines in trans-disciplinary and cross-cultural ways. They may include social systems theory, sociology, business administration, management information science, organization science, computational mathematical organization theory, economics, evolutionary economics, international political science, jurisprudence, policy science, socioinformation studies, cognitive science, artificial intelligence, complex adaptive systems theory, philosophy of science, and other related disciplines. In addition, this series will promote translational systems science as a means of scientific research that facilitates the translation of findings from basic science to practical applications, and vice versa.

We believe that this book series should advance a new frontier in systems sciences by presenting theoretical and conceptual frameworks, as well as theories for design and application, for twenty-first-century socioeconomic systems in a translational and trans-disciplinary context.

Kyoichi Kijima

Editor

Service Systems Science

 Springer

Editor
Kyoichi Kijima
Graduate School of Decision Science and Technology
Tokyo Institute of Technology
Tokyo, Japan

ISSN 2197-8832   ISSN 2197-8840 (electronic)
ISBN 978-4-431-54266-7   ISBN 978-4-431-54267-4 (eBook)
DOI 10.1007/978-4-431-54267-4
Springer Tokyo Heidelberg New York Dordrecht London

Library of Congress Control Number: 2014949896

Preface

The present volume illustrates a rich and promising research field in service, service systems sciences, which approaches service by combining and fusing two strands of sciences: the science of service systems and systems sciences of service. They overlap with a complementary emphasis.

Although ideas of service are not new at all, it is absolutely necessary now for us to cultivate a new frontier of service research. Indeed, the scale, complexity, and interdependence of today's service systems have been driven to an unprecedented level by globalization, demographic changes, and technology developments. The rising significance of service systems implies that service innovation is now a major challenge to practitioners in business and government as well as to academics in education and research. A better understanding of service systems is definitely required.

Many individual strands of knowledge and expertise related to service systems exist, but they often lie in unconnected silos. However, these no longer reflect the reality of interconnected economic activities. For example, manufacturers of engineering products adopt service-oriented business models while health care providers learn lessons from modern manufacturing operations. Indeed, there are wide gaps in our knowledge and skills across silos.

In response, service science, management, and engineering (SSME), or in short, service science, has emerged during the past decade as a transdisciplinary research field that aims to clarify, analyze, and design the structure and process of service systems.

Service science is strongly motivated to bridge the gaps by providing the science of service systems. Its vision is to discover the underlying logic of complex service systems and to establish a common language and shared frameworks for service innovation. To this end, a transdisciplinary approach is explicitly employed for research and education in service systems.

Service science puts the emphasis on commonalities and interdependencies between goods and services rather than on differences. Service science defines service as general, as value co-creation interactions among entities through which various values including social, economic, cultural, and even emotional values emerge. Service-dominant (S-D) logic, the main underpinning logic of service science, maintains that the roles of providers and customers are not distinct but rather are both just symmetric resource integrators for the co-creation of value.

Service science defines the service system as dynamic configurations of people, technologies, organizations, and shared information that create and deliver value to customers, providers, and other stakeholders. Many service systems also have changed, from the supplier value chain to the value network of all stakeholders. Furthermore, because value co-creation interactions between entities are modernizing rapidly primarily because of information technology, service science especially pays attention to modern value co-creation mechanisms based on a growing repertoire of IT-enabled business models and approaches.

To deal with complexity, interactions, and the network of, in, and among service systems, we need to take a more systemic view. "Complexity" is derived from the Latin verb *complecti*, meaning "to twine together", while the noun *complexus* means "network". The word "system" comes from the Greek *systema*, which means "a whole composed of many parts". Hence, these words and ideas are themselves closely related and their interdependency is evident.

"Systems sciences" defines a system as a whole composed of parts and then focuses on investigation of how and what properties emerge from interactions and the interrelationship among the parts. Because systems sciences offers a way of thinking in relationships and interaction and theories and models to address complexity, it is legitimate to develop systems sciences of service by explicitly focusing on systemic properties of service and service systems. Dr. Jim Spohrer, one of the advocates of service science, maintains that service science itself is a specialization of systems sciences that seeks to provide an evolutionary account of service system entities and their increasingly sophisticated value in co-creation interactions.

As a volume of the Translational Systems Science series, this book emphasizes, in particular, a translational systems sciences perspective when the authors are approaching service, service systems, and service innovation. Indeed, the book employs systems sciences as a common framework or common language not only to approach service in a holistic way but also to take a transdisciplinary approach aiming to explain, analyze, design, and support service systems and their evolution.

The editor and his group have organized International Service Systems Science Workshops and Symposia at the Tokyo Institute of Technology annually since 2008 to discuss, communicate, and share the cutting edge of knowledge and experiences about service systems science with pioneering researchers and practitioners from North America, Europe, and Asia. All of the contributors to the present volume have attended the workshops at least once so that they have contributed their chapter from the perspective of sharing the basic idea of service systems science.

The book is divided into two parts: Part I, "Service Systems Research Perspectives", and Part II, "Service Systems Practice".

In Part I there are six chapters. In Chap. 1, "Social Value: A Service Science Perspective", Jim Spohrer, Haluk Demirkan, and Kelly Lyons analyze the concept of social value from a service science perspective, assuming that social value is of great interest to governments, foundations, non-profits, and corporate social responsibility organizations, and is a central focus of many policymakers.

A bridging framework for social value and individual value is presented along with some future research directions.

In Chap. 2, "Translational and Trans-disciplinary Approach to Service Systems", Kyoichi Kijima examines research schemes of service systems science from a translational systems sciences perspective. The author illustrates service systems science by emphasizing a translational approach, where processes from logic to theory and modeling are connected all the way through actual practice, and then introduces some typical reference models in service systems science including the value orchestration platform model.

In Chap. 3, "Service Artifacts as Co-creation Boundary Objects in Digital Platforms", Anssi Smedlund and Ville Eloranta introduce the concept of service artifacts, which are boundary objects created by the digital platform owner that engage the end user and facilitate the knowledge processes. The authors discuss service artifacts from the point of view of service dominant (S-D) logic and identify three categories of those artifacts. They also present examples of different types of service artifacts to illustrate their conceptual findings.

Chap. 4, "Four Axiomatic Requirements for Service Systems Research", by David Reynolds and Irene CL Ng, emphasizes the relevance of the application of a systems science perspective to service. The authors synthesize the developments in service systems research so far with the hope of clarifying some of the key concepts, and they explore some of the insights gained from what is rapidly becoming a well-developed body of literature. They state that service systems should be the "basic abstraction" of service science research and argue for four axioms that are necessary to advance knowledge in the domain of service systems.

In Chap. 5, "Social Innovations—Manifested in New Services and in New System Level Interactions", Marja Toivonen builds bridges to combine the perspectives of service, social, and system innovations based on the state of the art in research. She begins by opening up the concept and central topics of social innovation, followed by analysis with a review of user-driven innovation and open innovation.

In Chap. 6, "The Limitations of Logic and Science and Systemic Thinking—from the Science of Service Systems to the Art of Coexistence and Co-prosperity Systems", Takashi Maeno discusses the point that services are not simply an exchange of objects, acts, and money, but are, rather, complex acts with an exchange of psychological satisfaction and emotions. Then he points out the limitations of logic and science; namely, he argues that logic and science provide only a simplified model of the world, referring to the concepts of the uncertainty principle, the science of complex systems, and even the self-referential nature of logic.

Part II, "Service Systems Practice", consists of four chapters. In the first three chapters, service systems practice in terms of the public sector, healthcare, and the private sector are discussed in that order. The final chapter uniquely argues that "meta" service systems practice focusing on service R&D program design.

In Chap. 7, "Canadian Governments' Reference Models", Roy Wiseman explains reference models of public service for government improvement. The author maintains that well-constructed reference models, consisting of a common framework and language to describe the business of government, can assist in "doing government better".

He concludes that focusing on how governments are achieving the outcomes through their programs and services moves the discussion to a new level.

In Chap. 8, "What Is 5S-KAIZEN: Asia–African Transnational and Translational Community of Practice in Value Co-creation of Health Services", Hiro Matsushita deals with movement of transferring 5S-KAIZEN to the value co-creation activities of the service sector including health care and medical services. Kaizen ("continuous improvement") although the 5 S's (sort, set, shine, standardize, and sustain) originates in the operational management methodology of the Japanese manufacturing sector. Based on participatory observation and action research, the author presents the movement from the perspectives of systems thinking and service systems management. Improving practices applied to African health services is also reviewed.

In Chap. 9, "Creating Information-Based Customer Value with Service Systems in Retailing", Timo Rintamäki and Lasse Mitronen illustrate how information-based value creation has implications for the way retailers design and manage their customer value propositions for competitive advantage. By analyzing data from Japan, the U.S.A, and Finland, they assert that understanding the roles of different channels in the individual stages of the customer experience provides valuable input for service system development.

Chap. 10, "Service R&D Program Design Aiming at Service Innovation", by Yuriko Sawatani and Yuko Fujigaki, is unique in that it discusses service R&D program design for promoting service innovation. Although most design activities are carried out at the planning phase, the authors point out that the execution-phase activities are more important to achieve program-level objectives by strengthening the linkage of R&D and innovation. These interactions between a program and projects create values not expected at the planning phase, so that program management has to encourage these post-value co-creation characteristics.

The editor believes that the present volume, as part of the Translational Systems Sciences series, will certainly contribute to promoting the science of service systems as well as systems sciences of service with insightful findings and implications based on a translational approach.

Tokyo, Japan   Kyoichi Kijima
31 May 2014

Contents

Part I Service Systems Research Perspectives

1 **Social Value: A Service Science Perspective** 3
   Jim Spohrer, Haluk Demirkan, and Kelly Lyons

2 **Translational and Trans-disciplinary Approach
   to Service Systems** 37
   Kyoichi Kijima

3 **Service Artifacts as Co-creation Boundary Objects
   in Digital Platforms** 55
   Anssi Smedlund and Ville Eloranta

4 **Four Axiomatic Requirements for Service Systems Research** 69
   David Reynolds and Irene CL Ng

5 **Social Innovations—Manifested in New Services
   and in New System Level Interactions** 83
   Marja Toivonen

6 **The Limitations of Logic and Science and Systemic
   Thinking—from the Science of Service Systems
   to the Art of Coexistence and Co-prosperity Systems** 97
   Takashi Maeno

Part II Service Systems Practice

7 **Canadian Governments Reference Models** 109
   Roy Wiseman

8 **What Is 5S-KAIZEN? Asian-African Transnational
   and Translational Community of Practice in Value
   Co-creation of Health Services** 129
   Hiro Matsushita

**9   Creating Information-Based Customer Value with Service
     Systems in Retailing**.. 145
     Timo Rintamäki and Lasse Mitronen

10 Service R&D Program Design Aiming at Service Innovation 163
     Yuriko Sawatani and Yuko Fujigaki

Index.. 175

Part I
Service Systems Research Perspectives

Chapter 1
Social Value: A Service Science Perspective

Jim Spohrer, Haluk Demirkan, and Kelly Lyons

Abstract This chapter provides an analysis of the concept of social value from a service science perspective. Social value is a concept of great interest to governments, foundations, nonprofits, and corporate social responsibility organizations and a central focus of many policymakers. Service science is an emerging transdiscipline for the (1) study of evolving service system entities and value co-creation phenomena and (2) pedagogy for the education of twenty-first-century T-shaped service innovators from all disciplines, sectors, and cultures who may become social value generators through cross functional engagements. A bridging framework for social value (as calculated by social entities) and individual value (as calculated by individual entities) is presented along with some future research directions.

Keywords Service science • Social entities • Social value • Transdiscipline • Value co-creation

1 Introduction: Motivations and Goals

What is social value? This chapter provides a definition and analysis of social value from a service science perspective. As we will come to see, social entities are collectives built up from individual entities in a nested, networked fashion. To begin, we consider an example of social value in the wild.

J. Spohrer
IBM Almaden Research Center, San Jose, CA, USA
e-mail: spohrer@us.ibm.com

H. Demirkan (✉)
University of Washington, Tacoma, WA, USA
e-mail: haluk@uw.edu

K. Lyons
University of Toronto, Ontario, M5S 3GS, Canada
e-mail: kelly.lyons@utoronto.ca

© Springer Japan 2015   3
K. Kijima (ed.), *Service Systems Science*, Translational Systems Sciences 2,
DOI 10.1007/978-4-431-54267-4_1

When geese and other migrating birds fly in V-formation, trailing birds benefit from the extra effort of the goose upfront or leader. The lead goose is efficient. As the leader becomes exhausted, a natural rotation of leadership occurs where the strongest and best positioned moves into the leadership role. How many generations of evolution of migratory birds were needed to create the genetic and behavioral patterns for this aerodynamic collaboration? What role did competition and predators play pruning the weaker trailing birds, allowing this unique form of collaboration to emerge?

In the evolution of human groups, a leader is also often efficient, who can make a way to make things run more quickly and smoothly when there are difficult choices. When no single right choice exists for individuals, leaders select a best choice—the choice of compliance, following or obeying the leader. A leader thinks of self and thinks of group well-being and often benefits most from the health and survival of the group. However, what about groups without leaders, how do they operate, and what are the pros and cons of leaders?

Let's try to answer our first question. What is social value? How can we compare the social value of leaders to the social value of such things as literacy or money? Does scale (population size) and level (*knowledge burden*[1]) matter a lot or a little? Social value is arguably created by any number of evolved or designed solutions to human challenges and opportunities. It includes social capital as well as the subjective aspects of well-being, such as their ability to participate in making decisions that affect them and others.

To answer these questions further a broad perspective on human history is needed. Service science, which is an emerging transdiscipline, provides one such broad perspective. A transdiscipline borrows from existing disciplines, without replacing them. Like any emerging science, service science provides a new way of thinking and talking about the world in terms of measurements on entities, interactions, and outcomes, but also adds diverse symbolic processes of valuing (Spohrer et al. 2011; Spohrer and Maglio 2010). Specifically, a service scientist seeks to measure the number and types of entities, interactions, and outcomes, in order to advance better methods, processes, and architectures for thinking, talking about, and shaping the world in terms of nested, networked service system entities and value co-creation phenomena, including their diverse processes of valuing (Spohrer et al. 2012). These concepts (service systems, value co-creation, processes of valuing) are rooted in a worldview known as service-dominant logic or SD logic (Vargo and Lusch 2004, 2008). In the parlance of SD logic, service systems are sometimes referred to as resource integrators and value co-creation is often exemplified in exchange. According to SD logic foundational premise (FP) 10 "Value is always uniquely and phenomenologically determined by the beneficiary." This premise

---

[1] The *knowledge burden* of a society (species) derives from the need to ensure that the next generation has the knowledge required to run all technological and institutional/organizational systems needed to maintain the quality of life of theirs and future generations and continue innovating, thus growing the burden (Jones 2005).

describes how the ultimate action in service exchange is in the processes of valuing is defined.

In fact, all entities, be they social entities (such as a nation, city, foundation, hospital, business, etc.) or individual entities (such as a person), each has implicit processes of valuing that they are sometimes able to make explicit and empirically evaluate against other explicit processes of valuing. Formal service system entities (as opposed to informal service system entities) can be ranked by the degree to which they are governed by written (symbolic) laws and evolve to increase the percentage of their processes that are explicit and symbolic. For example, early hunter-gatherer groups that existed before written language are a type of informal service system (social entity). However, today, modern nations have constitutions, written laws, regulations, and policies and create written reports evaluating their compliance, often further validated by external auditors. Modern service systems use information and communications technologies (ICTs) to augment their capabilities (Engelbart 1995). The augmentations create a reliance on technology (and other formal physical symbol systems), which add to the knowledge burden of society (Jones 2005). Growing knowledge with respect to ICT-related design, execution, storage, transmission, and reuse is creating opportunities for leading public and private sector organizations to configure service relationships that create extraordinary new value (Chesbrough and Spohrer 2006). More specifically, ICT provides the means to improve the efficiency, effectiveness, and innovativeness of organizations (Bardhan et al. 2010).

Often service science is framed in the context of business-to-business outsourcing services (Maglio et al. 2006; Spohrer et al. 2007). To address service design for social enterprises, refinements to the foundational concepts of service science have been proposed (Tracy and Lyons 2013). So like all early stage scientific communities, the language for talking about service systems and value co-creation phenomena continues to evolve, including approaches to incorporate the concept of social value into service science thinking (Spohrer 2009).

The emerging service science community greatly benefits from theoretical and empirical studies done by a growing number of service researchers (see Appendix). Empirical studies of the economic success of businesses that adopt SD logic have begun to appear (Ordanini and Parasuraman 2011). Some studies of social enterprises have also begun to appear (Tracy 2011; Tracy and Lyons 2013). These latter studies highlight noneconomic measures such as emotional value (e.g., reduced anxiety, increased motivation, increased self-esteem, a sense of empowerment or peace of mind) and social value (e.g., ethical sourcing) and suggest a great deal more research is needed.

The purpose of this chapter is to analyze the social value in terms of service science and provide research directions on what and how we can bridge social value and individual value.

In the next section, a short overview of social value from a conventional perspective is provided. Section 3 provides background on service science. Section 4 is an initial service science perspective on the concept of social value, and Section 5 concludes with future research directions.

2 Overview: Social Value

Psychologists have defined three kinds of individual orientations (cooperative, individualistic, and competitive) and used them as theoretical bases for many studies investigating the ways in which individuals approach, judge, and respond to others (Van Lange 1999). Van Lange (1999) conceptualizes social value orientation as that in which individuals maximize joint outcomes or maximize equality in outcomes, or both. Indeed, the leaders described in our opening section exhibit this social value orientation (Hakansson et al. 1982). However, societies are comprised of many people who have different orientations from competitive, to cooperative, to individualistic. The role of service entities such as nonprofits, governments, and funding agencies is not only to establish mechanisms to maximize joint outcomes and/or equality in outcomes but to be able to measure the resulting social value. This is a very challenging proposition when members of society have varying and conflicting systems of social values (Mulgan 2010).

The paper by Mulgan (2010) is one of the best short and practical overviews of social value from a conventional perspective. He highlights the fact that there is little agreement on what social value is even though funders, leaders of nonprofits, and policymakers are keen to measure and assess social value. The key obstacle to social value assessment is the misconception that social value is objective, fixed, and stable (Mulgan 2010). Instead, when social value is seen as subjective, changeable, and dynamic, we are more likely to be able to define appropriate social value metrics.

Mulgan (2010) notes that most people have an overly simplistic view of social, public, or civic value, which is roughly the value that national and regional social programs, foundations, nongovernment organizations (NGOs), social enterprises, and social ventures create. Over the last forty years, hundreds of competing methods for calculating social value have been created. Mulgan (2010) summarizes the pros and cons of the main approaches to measuring social value, including: cost/benefits, stated/revealed preferences, social return on investment, public/value-added assessments, adjusted quality of life/satisfaction, government accounting measures, and field-specific measures.

He also identifies several factors that explain why current measures of social value too often fail. First, value is in the eye of the beholder and cannot be assessed completely objectively. It is not possible to simply consider traditional economic principles such as supply and demand when social, psychological, and environmental factors come into play. Mulgan (2010) suggests that metrics and tools for measuring social value are useful if they help build markets, conversations, and negotiation in order to bridge between people and organizations that have needs and those that have solutions. It isn't sufficient to introduce clients and providers; an environment that encourages conversations and negotiations to take place must be created and nurtured. These environments can also help disenfranchised groups (such as homeless people, migrant workers, and people with mental illness) to have a voice in the market. These groups have social and economic needs but often do not have the resources or power to create a demand for suppliers of solutions and services.

A second factor contributing to problems with current social value metrics is the attempt to combine multiple perspectives (internal, external, and societal) into a single quantitative value. Rather than quantifying social value through a single number, Mulgan (2010) proposes a framework that can be used to rate proposals according to four dimensions concerning the concept of social value: strategic fit; potential outcomes or results; cost savings and economic effects; and risks associated with implementation of the proposal. In addition to rating the proposal on a scale of 0–5 along the four categories of value judgments, decision makers can include comments to support the ratings. Many of the judgments, ratings, and comments are made based on evidence and data available to the decision makers. The proposed framework also enables participants to include measures of the reliability of the evidence used to determine the ratings. The results of the social value judgments made using the framework are presented visually allowing multiple people to examine and question the measures. Over time, the ratings can be compared to actual social value assessments and can encourage consistency across decisions. The results can also be made public, keeping the decision-making and measurement process transparent and enabling communication across agencies.

Finally, Mulgan (2010) identifies the challenge of time as a factor contributing to the difficulty of measuring social value. For many social endeavors, value will not be realized until several years in the future and it is challenging to judge that future value against immediate costs. Using discounted rates as is done in the commercial world to value a given amount of money today according to the fact that it will be worth less in the future is not appropriate for governments and social organizations. Governments and social organizations give significant weight to the well-being of generations of society in the future so it not suitable to devalue the future social worth.

Convening stakeholders, providing a holistic view onto quantitative and qualitative points of view, making judgments (different values and processes of valuing), prioritizing issues, giving voice to the weakest in society (the disenfranchised), continuously listening and acting, managing complexity, and blending compassion with consequences are just some of the considerations. In many democracies, voters are usually willing to pay taxes for security (military, prisons, police force, and fire department), literacy (schools), infrastructure (roads, utilities), justice (courts), etc. However, other programs may be more controversial (e.g., sex education, drug treatment, homelessness, job training, housing, mental health therapy, animal rights, environmental protection). Part of the complexity is apportioning responsibilities across multiple levels—individuals, families, communities, cities, states, nations, and even continental regions such as the European Union. Another part of the complexity is the large number of cultural factors that come into play and across many hundreds of years of human history attitudes can vary dramatically.

Mulgan (2010) provides a state-of-the-art view on social value. Stepping back, a service science perspective on social value looks at how we got here. In broad strokes, a service science perspective recapitulates the evolution of our nested, networked ecology of service system entities—but before doing that let's introduce service science more fully.

3 Background: Service Science

Service science[2] draws on a great breadth of academic disciplines, without replacing them. How entities use knowledge to cocreate value is intimately tied to all disciplines, which can be thought of as societal fountains of knowledge. As disciplines create knowledge, which is woven into the fabric of society and becomes essential to maintain quality of life, that knowledge becomes part of the knowledge burden of that society (Jones 2005). What differentiates service science from all existing disciplines is that it is a transdiscipline, drawing on all and replacing none, with a unique focus on the evolution of service systems and value co-creation phenomena. Service science aspires to provide the breadth for T-shaped service innovators who have both depth and breadth of knowledge. Depth can be in any existing academic discipline, and appropriate breadth can improve communications, teamwork, and learning rates (IBM 2011). T-shaped innovators are able to bridge across disciplines applying their own knowledge depth to other knowledge areas.

A service science perspective, as we will see below, is a way of looking at the world through the lens of service science and SD logic. A physics perspective is a way of looking at the world and seeing a world of things made of atoms and forces, even though it is not possible for us to really *see* an atom. A computer science perspective is a way of looking at the world in terms of universal computing machines (e.g., physical symbol systems, Turing machines, etc.) and codes (e.g., symbols as both data and algorithms). An economics perspective is a way of looking at the world in terms of actors, supply and demand, externalities, and moral hazards. As we will see below, a service science perspective is a way of looking at the world in terms of an ecology of nested, networked service system entities and the value co-creation phenomena that interconnect them.

Human endeavors, such as sciences, build on philosophical foundations, and each science must first provide ontology (what exists and can be categorized and counted),[3] then epistemology (how we know and how others can replicate results), and finally praxeology (actions and how knowing matters or makes a difference)[4]. These three "ologies" explicitly or implicitly underlie all sciences; as humans, we seek knowledge of the world and of ourselves and then work to apply that knowledge through actions to create benefits for ourselves and others by changing aspects

---

[2] Service science is short for the IBM-originated name of service science, management, and engineering (SSME), since service science was originally conceived to be the broad part of T-shaped professionals that complements depth in any disciplinary area with breadth in SSME (IBM 2011). More recently service science has been referred to as short for SSME+D, adding design (Spohrer and Kwan 2009). Even more recently, service science has been referred to as short for SSME+DAP, adding design, art, and policy. The naming of a transdiscipline is especially challenging, and communities can debate pros and cons of names endlessly.

[3] New sciences may seem like stamp collecting or counting stamps to scientists in more mature sciences. For example, Lord Rutherford said, "All *science* is either physics or *stamp collecting.*" Service science is still at the stage of counting and categorizing types of entities, interactions, and outcomes.

[4] Thanks to Paul Lillrank (Aalto University, Finland) for this thought.

of what exists (e.g., service), in full awareness of our human sensory, cognitive, and motor limits—yet increasingly augmented by our technologies and organizations and augmented by scientifically and imaginatively derived knowledge, of both what is and what might be. However, all this knowing does create a knowledge burden which must be carefully managed (Jones 2005).

Quite simply, service is the application of knowledge for mutual benefits, and service innovations can scale the benefits of new knowledge globally and rapidly, but all this knowing does create a burden—including the burden of intergenerational transfer of knowledge.

Augmentation layers lead to the nested, networked nature of our world—specifically, as an ecology of service system entities. Value co-creation phenomena (service-for-service exchange) form the core of our human ecology (Hawley 1986). Value co-creation phenomena are also known as win–win or nonzero-sum games (Wright 2000). Competing for collaborators drives the evolution of markets and institutions and contributes to both their dynamism/stagnation and stability/instability (Friedman and McNeill 2013). Information technology, Internet of Things, big data, etc., are accelerating the ability of service systems to develop and continuously evolve and refine explicit symbolic processes of valuing, which further augment service system capabilities. Alfred North Whitehead, English mathematician, is quoted as saying: "Civilization advances by extending the number of important operations which we can perform without thinking of them" (Whitehead 1911, page 61). Augmentation layers, including technological and organizational augments, contribute to the nested, networked nature of our world and our knowledge burden (Angier 1998). Augmentation layers have many benefits, but they can also hide the extent of a society's knowledge burden.

The mature sciences of physics, chemistry, biology, and even computer science and economics can be used to tell a series of stories—overlapping and nested stories about our world and us. Physics describes the world in terms of matter, energy, space, and time, with fundamental forces well quantified across enormous scales to explain phenomena much smaller than atoms and much larger than galaxies. Physicists theorize and quantify to tell a story that stretches from before the big bang to beyond the end of time itself. Chemistry describes the world in terms of the elements, molecules, reactions, temperature, pressure, and volume. Geologists and climatologists, born of modern chemists, can tell the story of the birth and aging of our planet. Biology describes the world in terms of DNA, cells, and molecular machinery driven by diverse energy sources. Ecologists informed by modern biology tell the story of populations of diverse species shaping and being shaped by each other and their environments. Computer science describes the world in terms of physical symbol systems and other computation systems, codes, algorithms, and complexity. Cognitive scientists and neuroscientists are today working with computer scientists and others to propose stories of the birth of consciousness, communications, and culture in humans and prehuman species. Finally, economics describes the world in terms of supply, demand, externalities, principles, agents, moral hazards, and more. Economists theorize and quantify to tell the story of morals and markets, laws, and economies evolving over the course of human and even prehuman history and how the world can be in balance one moment and then go completely out of balance the next (Friedman 2008; Friedman and McNeill 2013).

Service science adds to these stories. Service science is enormously practical, as national economies and businesses measure an apparent growth in services in GDP (gross domestic product) and revenue. Getting better at service innovation is the practical purpose of service science. Service science is also academic, and like the academic discipline of ecology, it is an integrative and holistic transdiscipline drawing from (and someday perhaps adding to) other disciplines. While the basis of service is arguably division of labor and specialization, which leads to the proliferation of disciplinary, professional, and cultural silos, nevertheless service science, as an accumulating body of knowledge, can add some measure of breadth to the depth of specialists. In this sense, service science is holistic and inclusive, and every individual can add to her/his breadth as she/he adopts a service science perspective and learns more about how the overlapping stories of other sciences and disciplines fit together into a whole. The nested, networked nature of our world becomes more apparent. As service science emerges we can begin by seeing and counting service system entities in an evolving ecology, working to understand and make explicit their implicit processes of valuing and their value co-creation (stable change with many win–win experiences) and co-destruction (unstable change with many lose–lose experiences) interactions over their life spans. In a simple way, the goal of service science is to catalog and understand service systems and to apply that understanding to advancing our ability to design, improve, and scale service systems for practical business and societal purposes (Demirkan et al. 2009). The growth of service economies has broad implications for the well-being, society, operation of businesses, the creation of academic knowledge, the delivery of education, the implementation of government policies, and the pursuit of humanitarian causes.

In the remainder of this section, we more fully explain the emergence of service science as an effort to integrate the work of service researchers from many disciplines, while extending that research as well through a greater emphasis on service systems and value co-creation (see Appendix). We do this by summarizing historical service research and the relationship to the emerging service science community, both the academic discipline(s) and professional association(s), service-dominant logic, service science foundational concepts, service science foundational premises, a proposed research agenda for a science of service, and some proposed extensions to that research agenda, each in turn.

3.1 Service Research History and Community

Because many disciplines study service, there is a great need for a transdiscipline like service science. A more fully elaborated history of service research can be found in Spohrer and Maglio (2010). Over two-dozen academic disciplines now study service from their own unique disciplinary perspective, and not surprising, each has one or more definitions of *service* (Demirkan and Spohrer 2010).

Because many professional associations also have service-related Special Interest Groups (SIGs), journals, or conferences, because many nations and businesses have service innovation and service offering roadmaps, because many universities have or are starting service research centers, there is a great need for a transdiscipline like service science and an umbrella professional association like the International Society of Service Innovation Professionals (ISSIP), which promotes service innovation professional development, education, research, practice, and policy. ISSIP. org is an umbrella professional association that adds value to existing professional associations with service-related SIGs, conferences, and journals as a bridge.

Just as service science draws on without replacing existing academic disciplines, ISSIP draws on without replacing existing professional associations—by design. The ISSIP community is new but growing. Professional associations are a type of service system that can be designed and evolved, within a population of other professional associations competing for collaborators. In fact, professional associations are a kind of social service system with goals to maximize joint outcomes and quality of outcomes.

Why are many academic disciplines and many professional associations turning to service as an area of focus? First, since service is the application of knowledge to create mutual benefits, disciplines and professional associations are eager to show the way in which their body of knowledge can be applied to create real-world benefits. Sciences typically choose the path of creating engineered icons to demonstrate benefits (e.g., a bridge, a new material, a genetically enhanced plant), and arts typically choose the path of creating cultural icons to create benefits (e.g., a play, a song, a fashion). We remember icons because they inspire awe and create value for diverse beneficiaries. Engineering is good for creating certain types of realities, and arts are good for expressing as well as inspiring possible realities. Service expresses mutual benefit and borrows from business, engineering, arts, design, operations, psychology, and many others as to how those benefits are manifested depending on goals and needs of the service participants.

A summary of the main branches (i.e., economics, marketing, operations, engineering, computing, informatics, systems, organizations, law, etc.) that service science draws can be found in Appendix.

3.2 Service-Dominant Logic

For most people, the notion that goods have value seems obvious. Isn't that why we pay for them? However, service-dominant logic (SD logic) (Vargo and Lusch 2004) challenges us to change the way we think about goods, value, and more.

Value is not an intrinsic property of goods. For example, a physicist would have a hard time measuring the value of a good, although the mass and other physical properties could be measured. On the other hand, a lawyer could quickly assess the value of a good (e.g., property) a client lost access to through the negligent behavior of some other actor. Common sense tells us that the *price* one pays to own or lease goods can vary depending on market conditions and context. Common sense also tells us the price is not the value. A measure of the value runs straight into subjective customer experience. Ng (2012) talks about "worth" as a point-in-time decision about what one is willing to pay (the price) for something and value as a subjective, context-specific feeling of goodness at a later time. Customer knowledge and action can impact value realization (Auerswald 2012). For example, buying an exotic fruit, properly harvesting, transporting, storing, preparing, and then enjoying eating result in a positive value/feeling. On the other hand, if the fruit were to spoil and be thrown away, the result would be a negative value/feeling. These two examples demonstrate the way in which the customer's actions impact their experience.

SD logic is deeply rooted in a notion of value based on customer experience and outcomes, which is in turn rooted in customer knowledge and actions. By applying knowledge (e.g., eating the fruit in a timely manner versus letting it spoil) the customer cocreates value with the provider who made the fruit available to the customer just at the right time to maximize value for both of them. The customer may even store the fruit in a particular way to optimize the readiness of the fruit for a particular recipe. There is no end to how elaborate a customer's knowledge might be to realize an outcome. More and more, service innovators understand this view of active customers applying knowledge to cocreate value directly or indirectly with provider networks versus the view of passive consumers. Service innovators work to co-elevate both the provider and customer knowledge to realize more important and significant outcomes. Providers compete for customers, which is to say providers compete for collaborators.

SD logic makes an important distinction between *operand* resources and *operant* resources. The latter interact directly or indirectly to cocreate value (service-for-service exchange) and are also referred to as *actors* and *resource integrators*. Customer and provider actors are operant resources because they can apply knowledge to cocreate value. Operand resources, on the other hand, are the raw materials, tools, and information that can be used by the operant resources—if they, the actors, have the right knowledge to use them appropriately. Much of the service comes down to putting knowledge into action and then the processes of valuing the resulting experience and outcomes.

As we will see, in the parlance of service science, these actors (operant resource integrators) are called *service system entities* and can be people, businesses, universities, cities, nations, or any other entities capable of knowledge-intensive interactions based on value propositions and governed by rights and responsibilities (governance mechanisms). When operant resources interact directly or indirectly, it is both the experience and outcome of those interactions that concern service innovators.

For example, a car is not just a type of good that can be purchased and used, but a car is an operand resource that came to exist only through the interactions of many people and businesses over time, and these people and businesses are the operant resources with the capability to apply knowledge to create benefits for others and themselves. When you buy a car, you are really buying an unimaginably long series of service-for-service exchanges throughout history that led to the car. The money you use to buy a car summarizes an equally unimaginably long series of service-for-service exchanges.

To use a car as intended for transportation (to realize value) requires an operant resource (a driver) applying knowledge. Service is the application of knowledge for the benefit of others and self. To say this somewhat differently, operant resources apply knowledge to create benefits with other operant resources, directly or indirectly. According to SD logic, all goods and money are just operand resources that arise as a result of operant resources applying knowledge. So goods (or operand resources) have no intrinsic value. Instead, value resides in the experiences and outcomes of operant resources and is not something intrinsically within operand resources. While better explanations of applying knowledge and experiencing outcomes are necessary, suffice it to say SD logic provides a way to change the way we think and talk about the world and prepares us to think about service innovations more clearly—service innovators improve the way operant resources apply knowledge and experience outcomes. Service innovators design better games for the players (the operant resources); the goods (the operand resources) are props in the game. Better games raise the bar on outcomes. Some fundamental service innovations improve our ability to compete for collaborators, co-elevating our capabilities in the process.

With this background, the ten foundational premises of service-dominant logic as revised by Vargo and Lusch (2008) are:

SDL-FP1: Service Is the Fundamental Basis of Exchange
Implicit in SDL-FP1 is a definition of service as operant resources (actors) applying knowledge and skills for mutual benefits (value co-creation experiences and outcomes). Service-for-service exchange is the fundamental building block of all exchange ("I'll do this for you, if you do that for me" or more precisely "I'll put my knowledge into action for you, if you put your knowledge into action for me"). From a service science perspective, exchange is a type of knowledge-intensive value-proposition-based interaction between entities.

SDL-FP2: Indirect Exchange Masks the Fundamental Basis of Exchange
Implicit in SDL-FP2 is a definition of indirect exchange. For example, exchanges involving goods and/or money (so-called operand resources available to or derived from previous efforts of operant resources) obscure the fundamental service-for-service nature of exchange. The series of questions, "And where did that operand resource come from?" always lead back in human history to a person (operant resource) applying knowledge for mutual benefits, in some sort of service-for-service exchange. From a service science perspective, operant resources such as people and businesses have rights and responsibilities, but operand resources such as technology/things or information/ideas do not have rights and responsibilities.

SDL-FP3: Goods Are Distribution Mechanisms for Service Provision
Goods are operand resources. Well-designed goods incorporate a great deal of knowledge that may be the accumulation of the knowledge and practices of many people over many years.

SDL-FP4: Operant Resources Are the Fundamental Source of Competitive Advantage
Operant resources (e.g., people) can put knowledge into action and take responsibility for their actions. Certain people or businesses may possess unique knowledge or

capacity for safely taking on added responsibility (e.g., risk). Goods and information are operand resources, which in general are easier to copy than operant resources.

SDL-FP5: All Economies Are Service Economies

Implicit in SDL-FP5 is a definition of an economy. An economy is a population of operant resources with capabilities for exchange interactions. According to SDL-FP1, service is the fundamental basis of exchange. Therefore, all economies, hunter-gatherer, agricultural, extractive, information, etc., are based on service-for-service exchange between operant resources.

SDL-FP6: The Customer Is Always a Cocreator of Value

Implicit in SDL-FP6 is a definition of value co-creation. The customer is an operant resource and must apply knowledge in context to generate an experience and outcome. Win–win outcomes require both the customer and provider to realize benefits. It is worth noting that this concept confuses many people because they think of coproduction as a kind of physical work effort on the part of the customer. Work can be a direct physical collaboration (coproduction) or indirect cognitive/social coordination (co-creation). When I trust a cleaning service with the key to my house, and they trust me to pay them, we are co-creating value. When I stay at home, open the door, and get involved in cleaning my house with them, we are coproducing value. The value is in the experience and outcome, which can be derived from physical direct collaboration or trusted indirect coordination.

SDL-FP7: The Enterprise Cannot Deliver Value but Only Offer Value Propositions

Implicit in SDL-FP7 is a definition of value. Providers can assess the *cost* of service provision, but only the customer can assess the *value* of the experience and outcome. The customer can make a decision about the *worth* of an offer, based on the *price* and some mental simulation, expectation, or anticipation of the value. For example, even when an emergency response team is trying to rescue a person in peril, if that person does not want to be rescued, and does not comply or cooperate in the rescue, then it is more likely that the emergency response team will fail. Both the customer and the provider must agree to the value proposition and see the mutual benefit as well as the mutual responsibility. Win–win value propositions are at the heart of value co-creation interactions.

SDL-FP8: A Service-Centered View Is Inherently Customer Oriented and Relational

Provider value depends on customer value, which derives from experience and outcomes, and ability to apply knowledge. Win–win value propositions are at the heart of value co-creation interactions. Repeatable mutual benefits depend on mutual knowledge, trust, and coordination. Service innovators know that customer-to-customer interactions can scale value via word of mouth and platforms.

SDL-FP9: All Economic and Social Actors Are Resource Integrators

Implicit in SDL-FP9 is a definition of resource integrators. Operant resources are resource integrators, and they can apply knowledge to combine and configure (integrate) both other operant and operand resources. For example, a driver must know how to drive to benefit from a car, and a student must know how to read to benefit from a book (at least for the primary intended use). All economic and social actors apply knowledge to integrate resources. Resources can be divided into three categories: market-facing resources (available for purchase to own outright or for lease/contract), private non-market-facing resources (privileged access), and public non-market-facing resources (shared access). Service system entities are economic and social actors, which configure (or integrate) resources.

SDL-FP10: Value Is Always Uniquely and Phenomenologically Determined by the Beneficiary
Implicit in SDL-FP10 is a reference to value determination as a process, unique to each beneficiary. Therefore value determination is a process unique to each subject or a subjective process. However, this does not mean that the process is random or unknowable. Culture and education can shape the process of valuing. Value realization is more than a decision (anticipatory calculation of benefits or worth). Value realization is contextual, history dependent, and uniquely determined by the beneficiary, shaped by culture and education. Building models of these processes, and the way culture and education shape them, is essential to advancing service science. Furthermore, these models could provide a foundation for theoretical service science.

Vargo and Lusch are clear that these foundational premises are only a starting point, and they have worked with many others to continue the evolution of SD logic. For example, reducing the foundational premises to a smaller number of definitions and foundational axioms has been undertaken. Four of the foundational premises (SDL-FP1, SDL-FP6, SDL-FP9, and SDl-FP10) have been shown to be adequate for deriving the others (Vargo and Lusch 2008).

3.3 Service Science Foundational Concepts

The fundamental concepts of service science should facilitate the creation of a trading zone between many academic disciplines (Gorman 2010). A trading zone invites individuals from different backgrounds with different vocabularies to communicate, share ideas, and engage in mutually productive interactions. The value of the concepts below, versus some other fundamental set of concepts, is in giving individuals easier access to ideas from many different disciplines. One branch of the service science community, sometimes known as the SSME+PAD branch, identifies ten service science foundational concepts (SS-FC1-10: ecology, entities, interactions, outcomes, value propositions, governance mechanisms, resources, access rights, stakeholders, and measures), plus an additional eighteen foundational sub-concepts.

The concepts and sub-concepts should be general enough to allow many disciplines to contribute to the creation of service science and build a better understanding of service systems and value co-creation phenomena. We describe each in turn:

SS-FC1: Ecology
Service science borrows from ecology (populations) as much as from economics (price). Ecology as a discipline is the study of populations of entities (evolving,

competing, cooperating, etc.) and their relationship to each other and their environment. Ecology as a concept is quite general and can apply to atoms in stars (stellar nucleosynthesis), animals in a forest, or nested, networked service system entities. Measurement of the number and type of entities, interactions, and outcomes is fundamental to ecology (and the ontological foundations of a new science).

Service ecology as a concept provides the fundamental way of thinking more scientifically about service system entities—they exist as populations of entities (evolving, competing, cooperating, etc.) in relationship to each other and their environment and can be counted and classified. The population of service system entities forms the service ecology. Currently, the service ecology is based on just one foundational species, humans, which have evolved formal (written/computational symbol based) service system capabilities for assigning and externalizing the rights and responsibilities of service system entities as legal, economic, and social systems (Deacon 1997).

Order of magnitude observation: An interesting observation about the human service ecology is that as the population approaches ten billion people, the estimate of the total number of formal service system entities (with legal rights and responsibilities) is less than one hundred billion entities. The ten entities per person average may be tied to the structure of society. Each person plays a role in several other service system entities, for example, since over 50 % of the world's population lives in cities, most people are part of service systems for their nation, state, and city. If they have a job, they may be part of a business or social enterprise. It is also interesting that to a first level of approximation, most people (individuals) are nested ten levels deep in service systems ((1) world, (2) continental union, (3) nation, (4) state, (5) county-metro, (6) city, (7) district, (8) community, (9) street, (10) household). The rough order of magnitude relationship may have to do with human capabilities and limitations, as well as the sustainable knowledge burden level of augmentation with technology and governance mechanisms. The observation may also be related to the life span and sustainability of businesses of various scales. Some businesses are global and operate in nearly all nations, and other businesses are local to a street or community. Service is the application of knowledge for mutual benefits and transformative service innovations scale up the benefits of new knowledge globally and rapidly.

From a service science perspective, each individual human and many collectives are service system entities. Human families are hundreds of thousands of years old, cities only about 10,000 years old, universities that have survived to today are only about a 1,000 years old, and modern businesses with professional managers arguably just 100 years old. Looking at orders of magnitude across time, it is clear that the scale (population size) and level (knowledge burden) of the human ecology has grown dramatically. As population size increases, a society can take on a larger knowledge burden. In fact, there is archaeological evidence that as human populations become isolated and shrink (e.g., land bridges to islands disappear), the level of technological and other indicators of cultural complexity decreases (Kremer 1993).

A luxury cruise ship is a good example of a holistic service system (Motwani et al. 2012). A *holistic service system* (SS-FSC1) is a type of service system entity in the service ecology, such as a nation, state, city, university, hospital, cruise ship, and family/household, which provides whole service to the people inside the holistic service system (Spohrer et al. 2012). *Whole service* (SS-FSC2) refers to three categories of service capabilities necessary for quality of life of people inside service systems: flows (transportation, water/air, food/products, energy, information/communications), development (buildings/shelter, retail/hospitality/entertainment/culture, finance, health, education), and governance (rules that make competing for collaborators co-elevating) (Spohrer 2010). Holistic service systems can remain viable for some period of time, even if disconnected from all interactions with other external service systems for some period of time.

SS-FC2: Entities

Service system entities are the fundamental abstraction of service science (Maglio et al. 2009). A *formal service system entity* (SS-FSC3) is a legal, economic entity with rights and responsibilities codified in written laws. An *informal service system entity* (SS-FSC4) uses promises, morals, and reciprocity in place of contracts, written laws, and money (Friedman 2008). Mature, economically productive citizens of nations are formal service system entities with rights and responsibilities, but still operate as informal service systems when at home with their families. Children suing parents is an indication of the formal-informal boundary dispute/redefinition in progress.

Entity capabilities (SS-FSC5) *and constraints* (SS-FSC6) change over time. Capabilities and constraints impact the ability of entities to compete for collaborators and succeed at co-elevating forms of value co-creation. Human service system entity capabilities include physical, cognitive, and social capacity for work, including the ability to augment capabilities with technology and governance mechanisms. Human service system entity constraints include finite life span, finite learning rates (bounded rationality), and finite social networks, though augmentations change these constraints, while introducing a knowledge burden (Simon 1996; Jones 2005). Capabilities and constraints also include socially constructed rights and responsibilities, discussed below.

Entity identities (SS-FSC7) *and reputations* (SS-FSC8) change over time. Identity and reputation impact the ability of entities to compete for collaborators and their ability to succeed at co-elevating forms of value co-creation. Business service system entity identities and reputations contribute to brand and word-of-mouth marketing. National service system entity identities and reputations contribute to emigration, international student, and tourism rates. Individual human service system entity identities and reputations contribute to credit ratings and social network followers.

SS-FC3: Interactions

Measuring the number and types of interactions between service system entities is complex. *Service system entity interactions* can be well designed or spontaneous and then well or poorly executed. Also, interactions can be service interactions or non-service interactions. Service interactions are either value proposition based or governance mechanism based. Interactions that are value proposition based form networks that are both internal and external to the service system.

SS-FC4: Outcomes

Measuring the number and types of outcomes when service systems interact is complex; nevertheless, a few first-order simplifications can be made. For example, in the case of two entities interacting, a simple four-outcome model is: win–win, lose–win, win–lose, or lose–lose. Of these, only the win–win outcome is a service interaction with mutual benefits realized; nevertheless, in a nested, networked ecology of entities, even win–lose outcomes can serve a higher purpose. Beyond mutual benefits between two entities, when considering social value, we can also take into account benefits to the service ecology, other entities, and the community. In a study by Tracy and Lyons (2013) of social enterprises as service systems, social value was found to include benefits received by an entity indirectly as a result of a service interaction; that is, in social enterprises, value is realized when the client and provider interact which in turn results in social value being realized by additional entities that are not directly involved in that interaction. The entity realizing the indirect social value is sometimes a physical entity (e.g., a community or the environment) and sometimes conceptual (e.g., culture). Tracy and Lyons (2013) suggest that the social value that is realized in these contexts is value creation at a higher level.

Entities evolve in order to transform zero-sum games (competitions) that have winners and losers into larger non-zero-sum games (collaborations) in which every entity wins, creating an incentive to participate. This blended use of competition and collaboration to improve capabilities of entities is at the heart of value co-creation interactions and outcomes.

For example, the US National Football League has a series of weekly competitions (win–lose) and an annual draft that helps maintain competitive parity. This type of governance (system of rules or game) helps to keep the weekly games (win–lose) exciting and maximize fan interest and engagement, increasing revenue for teams, players, their management, and owners (win–win). Chess rankings pit near competitive parity players against each other (win–lose), making it hard to predict winners, but creating opportunities for incremental learning and improvement to get to the next ranking level (win–win).

For another example, the design of the European Union (EU), which won the Noble Peace Prize in 2012, created a continental scale service system entity (see the order of magnitude observation under SS-FC1 above) with component service system entities (nations). The design of the EU is an example of blended competition and collaboration to enhance capabilities, or value co-creation, intended to make the EU more competitive on the global stage and improve the quality of life in all component nations.

Interact-Serve-Propose-Agree-Realize (ISPAR) (SS-FSC9) is an elaboration of the simple four-outcome model with ten outcomes (Maglio et al. 2009). ISPAR includes both service and non-service interactions. Non-service interactions can either be welcomed or not welcomed, legal or not legal, and result in justice or not justice. Service interactions may not be realized if the proposal is not understood, or if it is not agreed to. Even if the proposal is understood and agreed to, the result may not be realized, and this can lead to a dispute or no dispute, which can be resolved or not resolved to the satisfaction of both entities.

Recovery (SS-FSC10) is a foundational sub-concept of great importance, especially when a series of outcomes is expected between entities over time and disputes arise when some outcomes are not realized to the mutual satisfaction of entities. Studies indicate that when a provider recovers well from a service failure, it can create a higher level of trust and loyalty with customers than if no failures had occurred (Magnini et al. 2007). Of course, this finding has many implications, if the motivation in optimizing provider value is seen as customer manipulation. Loyalty programs that provide enhanced benefits to customers, even when a failure has not occurred, can have a similar impact. Customer lifetime value-informed investment strategies also create enhanced outcomes that pay off over the lifetime of interactions and outcomes (Rust et al. 2000).

SS-FC5: Value Propositions

Value propositions are offers to play nonzero-sum games and are at the heart of competing for collaborators. Often the value to the provider of the offer is hidden and not accessible to the intended customer. In some cases the provider outcome is not hidden. For example, in so-called ultimatum games, a player may refuse benefits, if a culturally determined sense of fairness in allocation is not realized (Spohrer, Anderson, Pass, Agre 2009; Wright 2000). In the cases where provided benefit is unknown, the customer will evaluate the value proposition relative to similar offers from the provider's competitors. If no similar offers exist to choose from, then the offer can be viewed as competing at a higher level for the attention, time, and other resources of the customer.

The science of value proposition design is evolving rapidly (Spohrer, Anderson, Pass, Agre 2009; Wright 2000). The essential considerations include models of other stakeholders' capabilities, limitations, and processes of valuing. A customer must understand an offer, agree to the offer, and then contribute (cocreate, coproduce) to realize the benefits of the offer. The more sophisticated the offer, the greater the customer's capabilities must be to understand, agreed to, and realize the outcome. Stakeholders include the provider, customer, competitors, and authority. Competitors may include the customer (self-service), legitimate competitors (abiding by the same laws and constraints), non-legitimate competitors (criminals, black or gray markets), and authority (government or social sector programs) and even online or other competitors who can play by different rules.

Worth (SS-FSC11) *and value* (SS-FSC12) are different concepts (Ng 2012). Worth is a point-in-time decision about what an entity is willing to pay for some anticipated future value. Value is an experience of goodness that is contextualized by an individual. Individuals will evaluate if an offer (value proposition) is worth it and if an offer (value proposition) became realized as anticipated or provides superior value.

Risk (SS-FSC13) *and reward* (SS-FSC14) are unknowable in advance and so must be estimated (Adams 2000). Every offer (value proposition) has associated risks and rewards that may be hard to anticipate and estimate. Some entities have greater risk tolerance than other entities.

Processes of valuing (SS-FSC15) *and deciding* (SS-FSC16) are ultimately at the heart of service science. If we had a perfect model of our own processes of valuing and that of all other service system entities, and we had perfect data about the world and unlimited computational capabilities, then the science of service could become more objective. Advances in cognitive science and the brain sciences do in fact help researchers build betters models of processes of valuing and deciding, and increasing computational capabilities can help certain well-scoped systems operate more optimally. Processes of valuing generate options and rank them. IBM's Watson supercomputer, known for its prowess in outscoring the top human Jeopardy! winners and creating diagnostic options for doctors to consider, is an example of a system with algorithmic processes of valuing (Ferrucci et al. 2010). Processes of deciding are tied to action. A decision with no action is not a true decision. Risk tolerance often prevents individuals from taking action even when processes of valuing suggest great potential rewards for certain options. It may be worth it to have others take the actions, but principal-agent problems may then arise, creating a different type of risk. Perhaps the order of magnitude observation (see above SS-FC1) combined with better governance mechanisms (SS-FC6) may offer a solution to many types of principal-agent problems, thus advancing the practice of value proposition design.

SS-FC6: Governance Mechanisms

Governance mechanisms are based on a system of rules or laws that constrain entity interactions, with coercive power. Formal service system entities (SS-FSC1) exist as formal entities because of their rights and the power of an authority service system entity to recognize, protect, and uphold those rights. Smart machines do not yet have rights. Businesses do have rights because of laws. Both laws and technologies contribute to the knowledge burden of society. A nation without coercive powers would have to exist based purely on voluntary value propositions and no such nation exists. The weakest form of coercive power is banishment, or cessation of existence *here*. The strongest form of coercive power is death of individual, family, and species with permanent erasure of historical mentions or cessation of existence *everywhere for all time*.

Only one set of service system entities legitimately retains rights to coercive value propositions that can threaten the fundamental rights including the right of refusal and right to exist, and that is government authorities. Criminal service system entities also use coercion, but they operate outside of national and international laws. All other service system entities are restricted to voluntary value propositions and use coercive value proposition only in criminal or private/nonpublic situations.

Rights (SS-FSC17) *and responsibilities* (SS-FSC18) go hand in hand. Rights are a privilege earned through responsible actions. Unless a service system has the capability to understand the responsibilities that accompany rights, they cannot enjoy those rights. Young children, debilitated elderly, and other individuals' cognitive or mental impairments may have restricted rights, because of their limited cognitive capacities.

SS-FC7: Resources

Resources can exist in four types: people; technology; organizations; and information. People and organizations are operant resources (actors), and technology and information are operand resources (used by actors). People augment themselves with technology and organizations to increase their capabilities and overcome constraints. This augmentation can positively impact quality of life, but can also introduce a significant knowledge burden on society. The size of the knowledge burden is reflected in the quantity of shared information. Shared information includes language, laws, measures, and much more.

Resources exist in context, as either as physical or not physical, and with rights or without rights. In the context known as the real world, people are an example of a physical resource with rights and businesses are an example of a nonphysical resource with rights. Even though a business may have buildings, or component physical resources, no physical component is essential to a business, and a business can stay in existence with none of the original buildings or people that were originally part of it. However, the body of a person is an essential part of that person, and so a person is a physical, with-rights resource, even though a person as a service system entity includes far more than just the body of the individual person. A person as a service system entity is a much larger resource constellation or configuration of component resources. For example, my car and house are component resources with the service system entity, which makes me up as an individual.

SS-FC8: Access Rights

Access rights include owned outright, leased or contracted, shared access, and privileged access. Owning property versus leasing property comes with different rights and responsibilities. Similarly, shared access resources, such as roads and the air we breathe, come with different rights and responsibilities, compared to privileged access resources, such as one's own thoughts or family members.

SS-FC9: Stakeholder Roles

Stakeholder roles include customer, provider, authority, and competitor. An employee may be viewed as all of a provider to a business, a customer of the business' benefits program, an authority governing and resolving disputes associated with certain business processes, and a competitor of another employee interested in the same organizational role. Service system entities are at once customer, provider, competitor, and authority, depending on the perspective. When considering social value, it is also useful to consider the community surrounding the service system as a stakeholder (Tracy 2011; Tracy and Lyons 2013).

SS-FC10: Measures

Measures include quality, productivity, compliance, and innovativeness. Many other measures and key performance indicators can be associated with service system entities or processes in which an entity participates. Measures allow ranking of service system entities. For example, universities (as service system entities) may be ranked based on the starting salaries of their graduates. Holistic service system entities may be ranked based on innovativeness, equity (competitive parity), sustainability, and resilience. Social organizations can be measured by resulting social value to participating entities and the broader service ecology.

3.4   Service Science Foundational Premises (SS-FP)

Maglio and Spohrer (2013) have been evolving foundational premises for service science. Others linking the concept of viable systems to service systems are also working on foundational premises (Barile and Polese 2010; Boulding 1956). An extension and evolution under consideration is presented below:

SS-FP1: All Viable Service System Entities Dynamically Configure Four Types of Resources: People, Technologies, Organizations, and Information

Put another way, a service system that cannot dynamically configure resources is not viable. The application of knowledge to dynamically configure access to resources for mutual benefits is a fundamental capability of service system entities, and often access to resources (rights and responsibilities) must be earned. For example, earning a driver's license is an earned right that requires demonstrating capabilities and taking on additional responsibilities. Earning and using a driver's license in society requires access to people (e.g., driving test certifier), technology (e.g., a car), organizations (e.g., Department of Motor Vehicles), and information (e.g., rules of the road booklet and test). For example, setting up a business is an earned right that requires capabilities and taking on additional responsibilities—people (e.g., hiring employees), technology (e.g., equipment or environmental resources used in the business), organizations (e.g., working with suppliers), and information (e.g., submitting tax forms on time).

SS-FP2: All Viable Service System Entities Compute Value Given the Concerns of Multiple Stakeholders, Including Customer, Provider, Authority, and Competitor

Put another way, a service system that cannot compute value given the concerns of multiple stakeholders is not viable. For example, a business must offer something of value to customers, maintain relationships with supply chain organizations (providers), obey any regulations that apply to the business (authority), and in the long run outperform competitors.

SS-FP3: All Viable Service System Entities Reconfigure Access Rights Associated with Resources by Mutually Agreed-to Value Propositions or Governance Mechanisms

SDL-FP9 states that all social and economic actors are resource integrators. All economic and social actors apply knowledge to integrate resources. Resources can be divided into three categories: market-facing resources (available for purchase to own outright or for lease/contract), private non-market-facing resources (privileged access), and public non-market-facing resources (shared access). Access rights fall into four categories: own-outright, lease/contract, privileged access, and shared access. Ensuring that nested entities have protected rights and comply with responsibilities is work performed by a governing authority.

SS-FP4: All Viable Service System Entities Compute and Coordinate Actions with Others Through Symbolic Processes of Valuing and Symbolic Processes of Communicating

Written laws and contracts are a relatively new innovation in human history. Computers, spreadsheets, expert decision support systems, and electronic trading systems are even newer innovations. The transition from purely informal promises (moral codes) to formal contracts (legal codes) speaks to the evolution of service systems from primarily informal to increasingly formal. Viewed from the perspective of computer science, artificial intelligence, and organization theory, people and organizations can be modeled as a type of physical symbol system (Newell and Simon 1976; March and Simon 1958). Technological and organizational augmentation layers contribute to the nested, networked nature of the service system ecology (Arthur 2009).

SS-FP5: All Viable Service System Entities Interact to Create Ten Types of Outcomes, Spanning Value Co-creation and Value Co-destruction

ISPAR (SS-FSC9) is an elaboration of the simple four-outcome model (win–win, lose–win, win–lose, or lose–lose) to ten outcomes (Maglio et al. 2009). As articulated in SS-FSC9, ISPAR includes both service and non-service interactions each resulting in one of several outcomes.

SS-FP6: All Viable Service System Entities Learn

If service systems can only apply knowledge in fixed patterns, they will not be able to compete with service systems that learn, adapt, and change to become more competitive. According to the Abstract-Entity-Interaction-Outcome-Universals (AEIOU) theory, service system entities perform four primitive economic activities (production, distribution, consumption, recycling) jointly or separately in time and space (Spohrer and Demirkan 2013). Service systems are complex adaptive systems made up of people, and people are complex and adaptive themselves. Service system entity interactions often exhibit learning curves, or efficiency improvements based on number of interactions (Spohrer, Maglio, Bailey, Gruhl 2007).

3.5 Proposed Research Agenda for a Science of Service

The service research community has taken some steps to define a research agenda and establish research priorities to advance the science of service (Ostrom et al. 2010). The ten priorities include strategic, development, and execution priorities from a managerial (marketing and operations) perspective and one pervasive priority from an engineering (computing) perspective; each priority is described briefly below:

SS-RP1: Strategic Priority: Fostering Service Infusion and Growth

This research priority deals with the ability of organizations to create and improve service offerings to grow. Changing culture (customer focus, service logic, servitization), strategy, business models (outcome-based), and portfolio management are important research topics related to this priority.

SS-RP2: Strategic Priority: Improving Well-Being Through Transformative Service

This research priority deals with the ability of governments and social enterprises to create and improve service offerings to improve quality of life for citizens and the disenfranchised. Social welfare (health, education), environment (sustainability, green), democratization (open data, transparency), urbanization (smarter systems), and bottom-of-pyramid issues are important research topics related to this priority.

SS-RP3: Strategic Priority: Creating and Maintaining a Service Culture

This research priority deals with ability of organizations to create and maintain a service culture. Human resources (hiring, training, and incentives), globalization (diversity), mind-set (values), and learning (adaptation) are important research topics related to this priority.

SS-RP4: Development Priority: Stimulating Service Innovation

This broad research priority deals with the ability of organizations to innovate to compete. Drivers (globalization, automation), types (incremental, radical), roles and sources (employees, customers, supplier, research, managers, universities), methods (design, arts, creativity), tools (modeling, simulation), and policy (investment, measurement) are important research topics related to this priority.

SS-RP5: Development Priority: Enhancing Service Design

This research priority deals with the ability of organizations to design better customer experiences and outcomes. Thinking (design, systems, processes), arts (performance, visual), challenges (economic cycles, cultural variations, market segments), and methods (collaborative, crowdsourcing) are important research topics related to this priority.

SS-RP6: Development Priority: Optimizing Service Networks and Value Chains

This research priority deals with the ability of networks of organizations to optimize collective performance. Supply chain, outsourcing, value migration, interorganizational governance, globalization, productivity, and optimization algorithms are important research topics related to this priority.

SS-RA7: Execution Priority: Effective Branding and Selling of Services

This research priority deals with the ability of organizations to establish brands to enhance sales. Social media, word of mouth, multichannel, consistency, assessment of brand value, sales force, and employee training are important research topics related to this priority.

SS-RA8: Execution Priority: Enhancing Service Experience Through Co-creation

This research priority deals with the ability of organizations to fully utilize co-creation. Sharing (responsibilities, work effort, risks, rewards, information, and property rights), role of actors (employee, customer, and manager), role of technology (channels, complexity), customer community management, recovery, and loyalty are important research topics related to this priority

SS-RA9: Execution Priority: Measuring and Optimizing Value of Service
This research priority deals with the ability of organizations to measure and optimize processes. Self-service technologies, return on investment, instrumentation, estimation, standards, portfolio management, and optimization algorithms are important research topics related to this priority.

SS-RA10: Pervasive Force: Leveraging Technology to Advance Service
This research priority deals with the ability of organizations to keep up with and incorporate disruptive technologies into service operations and to use advanced technologies to improve service offerings and customer experience. Platforms (smart phones, cloud computing, smart systems, web services, service-oriented architectures), accelerating change (business models, acquisitions), self-service technologies, real-time decision-making (cognitive computing, stream computing), security, privacy, and biometrics are important topics related to this priority.

Translating these priorities into a set of grand challenge research questions for service science remains to be done, though there have been some tentative efforts in this direction (Tang 2012).

3.6   Proposed Extensions to the Research Agenda

The ten research priorities in the previous section can be seen as priorities aimed at impacting practice with largely managerial and engineering implications. We propose three other priorities aimed at education, policy (tooling), and theory.

SS-RA11: Extension Education Priority: Curriculum
Creating curriculum and best practices for teaching and learning service science is an additional research priority. A curriculum that is designed to create T-shaped service innovators with depth and breadth, who have interactional expertise across disciplines, sectors, and cultures, is being requested by leading employers, to improve innovativeness, teamwork, and learning rates (IBM 2011).

Since service science is a transdiscipline and borrows from so many other disciplines, one interesting proposal for service science curriculum is optimizing the recapitulation of history from a technological and governance perspective (Spohrer 2012). Rapidly rebuilding societal infrastructure and institutions, without the many twists and turns of history, might allow for a compressed, integrated, holistic curriculum. This is also possibly an approach to reducing the knowledge burden, without reducing quality-of-life measures. Ultimately, service innovations, because they depend increasingly on symbolic knowledge and symbolic processes of valuing, must address the rising knowledge burden and the intergenerational transfer of knowledge challenges.

SS-RA12: Extension Policy Priority: Global Simulation and Design Tool

Creating a global simulation and design tool for evaluating alternative governance mechanisms is an additional research priority. Modeling the nested, networked service ecology could also have a profound impact on teaching and learning service science, especially if appropriate pedagogical idealizations can be developed (Spohrer and Giuiusa 2012).

Based on the order of magnitude observation, there is a much larger market for individuals than cities, a larger market for cities than nations. The global simulation and design tool could be used to experiment with policies intended to improve competitive parity between regions at all order of magnitude scales, while increasing the speed innovations could spread globally.

SS-RA13: Extension Theory Priority: Foundations

To put service science on a more fundamental theoretical foundation, it might be a useful research priority to consider a nested, network service ecology based on something other than the human species. For example, a service ecology based on intelligent machines, with greatly extended life spans, much faster learning rates, and much larger and denser populations, might be useful for thinking about a service ecology in the limiting case, when constraints on the basic building block service system entity (individuals) are removed. Alternatively, a service ecology with a diversity of species with different physical, cognitive, and social constraints could open up new theoretical directions for service science. Some work on an AEIOU (Abstract-Entity-Interaction-Outcome-Universals) framework has begun and greatly elaborated this could be part of an expanded theoretical foundation for service science and other transdisciplines (Spohrer and Demirkan 2013).

Understanding and characterizing the fundamental constraints on a species is an important area of research for developing the theoretical foundations for service science. For example, humans have the following constraints:

1. Physical: finite life span
2. Cognitive: finite learning rate
3. Social: finite population size/density

In the last 200 years, life spans have extended, education levels have risen, and population size/density has increased. In complex service systems, as fundamental (weakest link) constraints are removed, other constraints emerge to dominate system performance (Ricketts 2012). The mapping of fundamental constraints for other types of service system entities has not been developed yet.

4 Contribution: Bridging Framework

In many ways a service science perspective on social value is loosely consistent with Mulgan (2010), specifically: convening stakeholders (trading zone), providing a holistic view onto quantitative and qualitative points of view (transdiscipline), making judgments and prioritizing issues (understanding different values and processes of valuing), giving voice to the weakest in society (the disenfranchised as stakeholders), and continuously listening and acting to manage complexity (knowledge burden awareness, T-shaped individuals).

From a service science perspective, the beneficiary uniquely determines value, and so the differences and similarities of the processes of valuing used by both individual and collective service system entities become of great research interest. What are the characteristics of the processes of valuing, used by a collective service system entity, such as a nation? For example, a collective entity may have a process of valuing that considers any of the following a benefit:

1. Improved interactions with other entities (e.g., win–win mechanisms)
2. Improved rankings relative to other entities (e.g., competing for collaborators)
3. Improved capabilities of sub-entities (e.g., voice for disenfranchised)
4. Reduced knowledge burden (e.g., simpler, greener energy sources or materials)

From a service science perspective, in the case of the above process of valuing, the benefits (social value) of leadership derive from improved governance mechanism interactions, the benefits (social value) of literacy derive from greater capabilities of sub-entities, and the benefits (social value) of money derive from improved value proposition-based interactions.

Tracy and Lyons (2013) found that value co-creation in the context of social enterprises goes beyond assessments of quality and price. In social enterprises, the beneficiary can be the customer as well as society or the community. Even customer perceptions of value go beyond quality and price to include assessments of social, emotional, and functional value. Tracy and Lyons (2013) report that social enterprises (not unlike social organizations such as nations) make use of complex hybrid value propositions which have both intrinsic and extrinsic notions of value.

Thus, from a service science perspective, defining social value becomes reformulated to the empirical task of making explicit the processes of valuing used by different types of collective service system entities.

Considering the SS-FCs in Sect. 3.3, we can depict the service ecology (SS-FC1), entities (SS-FC2), interactions (SS-FC3), outcomes (SS-FC4), value propositions (SS-FC5), and stakeholder roles (SS-FC9) in Fig. 1.1. Within a service ecology, when two entities (each with their own stakeholders) interact through a value proposition, outcomes are achieved for each of the entities.

In Fig. 1.2, we show how making the notion of social value explicit changes the relationships among the service science foundational concepts. First, the community stakeholder is made explicit. Second, in addition to value propositions associated with the interaction between two entities, there are value propositions with the broader ecology and community stakeholders. Finally, there can be outcomes to the community resulting from the interactions between two entities.

In spite of the synergies between social value and the service science foundational concepts and SD logic foundational premises, one of the ultimate challenges in defining social value from a service science perspective arises from SDL-FP10. If the beneficiary is not an individual, but all stakeholders, all the citizens of a

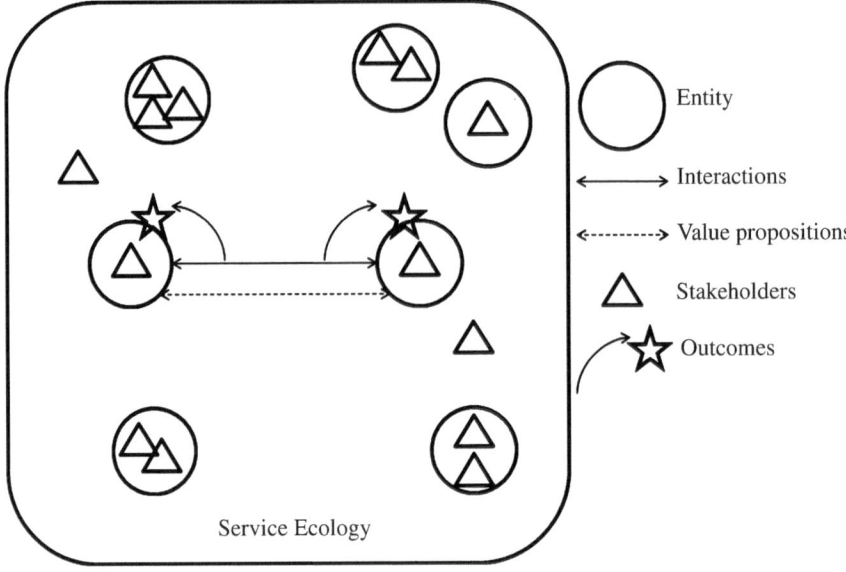

Fig. 1.1 Depicting service science foundational concepts

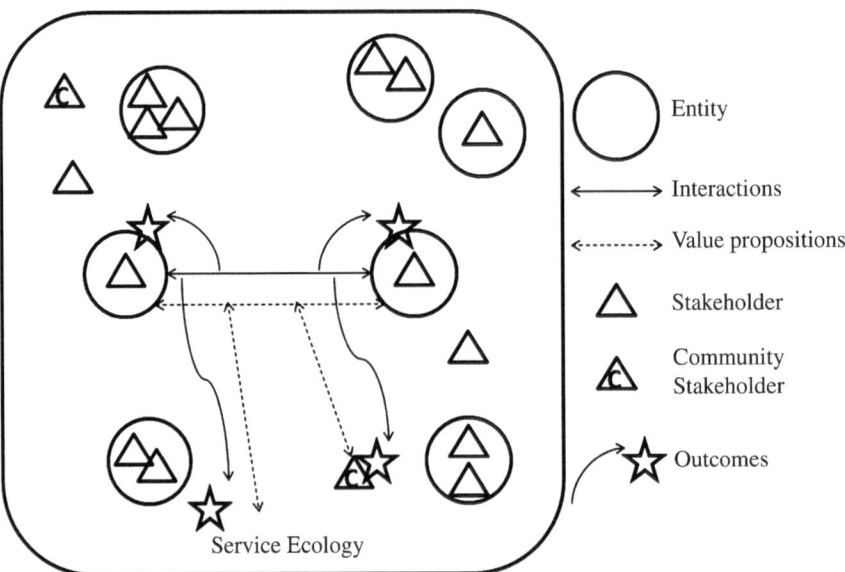

Fig. 1.2 Making social value explicit in service science concepts

nation, does that include criminals and revolutionaries seeking independence or other groups who are working outside the authority and governance mechanisms? The formal legal nature of service science becomes evident when asking these questions. Since formal service system entities seek to make explicit (e.g., symbolic and objectively operational) their processes of valuing, it is possible to estimate the social value from the perspective of any such formal service system entity. Can those entities revise their processes of valuing? Yes. Can those entities possess multiple competing processes of valuing? Yes. Can some processes of valuing be more easily operationalized than others? Yes. So in the end, what we are left with are multiple individual stakeholders with competing processes of valuing, which may be quite inconsistent and incomplete. In fact, some individuals who are part of the formal service system may not care enough or have enough capability to even have an opinion. Such is the complex nature of nonindividual formal service systems. Mechanisms for combining many individual perspectives exist, for example, electing a leader and adopting that individual's process of valuing as a surrogate for that of the electorate. Other mechanisms might include mechanisms for individuals to vote with their wallets, their time, or other resources, to provide those who care most about the issue to have a larger say in what is finally adopted and put into use. However, what about the weak, or disenfranchised, how does crowd funding (governance mechanism innovation) help them, when they have no financial resources?

The life span of any service system entity can be seen in terms of interactions and outcomes, but underlying those interactions are many instances of processes of valuing and decisions on how to act. Processes of valuing impact not only how we evaluate value propositions, but how we negotiate/agree and then work to realize the outcomes agreed to. How can we bridge individual processes of valuing with social entity processes of valuing? Both are often distributed cognition (Hutchins 1995), in the sense that even an individual's processes of valuing may include reaching out to ask the opinion of others or going online to research some alternatives. Processes of valuing are distributed in our cognition, and studies of decision-making when people are sleepy, hungry, emotional, or inebriated show cognitive effects, including delayed reaction times. When and what we eat and drink, when we sleep, when we get out of bed, what we do, how we react to things, all these and more are part of individual processes of valuing. What we decide is a separate process, but our individual processes of valuing create a list of candidates and rank orders them, even if it is only as simple as "Do It" or "Don't Do It." Unless our processes of deciding are based on something unusual, we probably select the top ranked choice from our processes of valuing. In social environments, one must weigh many aspects from multiple perspectives when making decisions. Rank orders have to take into account multiple perceptions of value.

Finally, over time service scientists working to understand and innovate social value must develop and apply relevant frameworks, theories, and models of social value. Ostrom (2009) proposed a specific relationship between frameworks, theories, and models that we adopt and extend. A framework provides shared language to describe real-world phenomena in terms of concepts and qualitative relationships that sharpen shared observations about what exists and how it came to exist (ontology). A theory provides rigor both in terms of measurement methods and empirically testable propositions to expand what is known and how it comes to be known and more efficient ways to arrive at and accumulate knowledge (epistemology). A modern model provides boundary conditions on a theory as well as a computational implementation that can be used to design, engineer, and manage new instantiated systems and realize benefits of theory-based knowledge constructs through appropriate real-world actions (praxeology). For example, the literature on social value from economic theories of social value has considered private value versus the social value of information, which potentially has practical consequences in the design of patent systems (Hirshleifer 1971). Also, we need to keep in mind that at the end of the day, we are debating about experiments to perform on ourselves. Nations and societies are like petri dishes.

5   Concluding Remarks: Future Directions

We live in a human-made ecology of nested, networked service system entities—people, families, businesses, universities, cities, states, nations, and more. Humans are unique in our ability to communicate, collaborate, compete, and realize shared dreams about the future, from start-up grand challenges (like building a social graph of the world) to national grand challenges (like landing a man on the moon), to scientific grand challenges (like mapping human DNA). Humans have evolved to compete for the cooperation of larger and larger groups of others. Many competitions are in fact mechanisms for cooperation in disguise, positively reinforcing rule-following compliance and punishing rule violations. Balancing competition and cooperation to accelerate learning and social benefits is fundamental.

The human ecology of nested, networked service system entities has already evolved through several technical infrastructure stages, remarkable in terms of energy, transportation, and communications, which enable great cities to emerge at an accelerating pace (Hawley 1986). Designing alternative viable futures for people in an age of rapidly increasing technical and organizational capabilities presents many challenges and opportunities. For example, policymakers understand that norms and laws must coevolve with technical capabilities created by engineers. Two important types of constraints shaping the evolution of service systems are the technical and environmental capabilities (infrastructure) and governance responsibilities (institutions). These two constraints interact with two other constraints, the education and skill levels (individuals) and quality-of-life aspirations of families (cultural information). Service is the application of knowledge for mutual benefits (value co-creation). Service innovations scale the benefits of new knowledge globally and rapidly. T-shaped professionals are professionals with depth and breadth of knowledge across academic disciplines, industry sectors, and regional cultures. T-shapes balance depth and breadth to optimize abilities to compete as individuals and collaborate in teams. Appropriate breadth has the potential to improve innovativeness, teamwork, and learning rates.

In this chapter, within the context of providing a service science perspective on social value, we presented a preliminary bridging framework for analyzing the historical evolution of service system entities to date and exploring the design space for alternative viable futures. Surprisingly, we argue that dealing with the knowledge burden of society, which helps people develop the skills to rapidly rebuild societal infrastructure and institutions along alternative possible historical pathways, may open up the largest design space for alternative viable futures. This chapter has implication for those in academics, industry, government, and the social sector interested in a more service-oriented view that balances past, present, and future possibilities.

A good ending point for an exploration of the concept of social value from a service science perspective is the quote from George Box saying that, "Essentially, all models are wrong, some are useful" (Box 1979, page 202). We believe that creating a trading zone for the development of service science as a transdiscipline, which borrows from disciplines without replacing them, is a useful and timely model. However, much work remains on multiple fronts to create more T-shaped service innovators, including advancing the practice, education, tooling/policy, and theoretical foundations for a science of service.

Acknowledgments Discussions with many colleagues at service science-related conferences around the world as well as email and social media interactions with ISSIP.org members globally have helped shape these ideas.

6   Appendix: Concepts Discipline, Researcher, etc.

Researchers from many disciplines have contributed to advancing service science and the study of service systems. Based on a sampling of publications (Spohrer 2013—http://service-science.info/archives/2708), some disciplinary branches are partially summarized in the table below.

Concept	Discipline	Researcher	Journal	Conference	Association
Stakeholder customer	Marketing	Rust	JSR, CACM	Frontiers	AMA, INFORMS, ASA
		Fisk	JSR	Frontiers, AMA SERVSIG	AMA
		Bitner	JSR, CACM	Frontiers	AMA
		Vargo	JM, JAMS	Frontiers	AMA
		Lusch	JM, JAMS	Frontiers	AMA
		Gronroos	JSR, JAMS	Frontiers, QUIS	FSSL
		Edvardsson	JAMS	Frontiers, QUIS	
		Gummesson	JBIM	Forum, QUIS	SSEBA, ISQA

(continued)

(continued)

Concept	Discipline	Researcher	Journal	Conference	Association
Stakeholder provider	Production operations	Sampson	JSR	POMS	POMS
	Operations management	Neely	OMR	Alliance	EOMA
		Davis	IBMSysJ, OMR	ArtSci	DSI, POMS
		Metters	DS	POMS	DSI, INFORMS, POMS
		Apte	POMS	POMS	POMS, DSI
	Operations research	Larson	JoSS		INFORMS
		Badinelli	JoSS	Forum	INFORMS,ISSIP
Stakeholder authority	Governance	Piciocchi	JoSS	Forum	ISSIP
		Bassano	JoSS	Forum	ISSIP
Stakeholder competitor	Strategy	Polese	JoSS	Forum	ASVSA
		Barile	JoSS	Forum	ASVSA
Resource people	Social sciences anthropology	Baba	CACM	HSSE	AAA NAPA
	Cognitive science	Glushko	JSR, IBMSysJ	Frontiers, HSSE	CSS, OASIS
	Human factors	Freund	HFEMSI	HSSE	HF&E, IIE, ISSIP
Resource technology	Industrial engineering	Rouse	IBMSysJ		IIE, INCOSE
	System engineering	Tien	JSSE		IEEE, NAE
		Berg	JSSE		IEEE, NAE
Resource information	Computer science	Spohrer	CACM, JAMS, Computer	Frontiers, HSSE, AMCIS	ACM, ISSIP
		Maglio	CACM, JAMS, Computer	HICSS	ACM
	Information systems	Alter	IBMSysJ	AMCIS	AIS, IFIP
		Demirkan	CACM, ECRA, JMIS, JSR	AMCIS, HICSS	AIS, ISSIP
		Kwan	IJISSS	AMCIS	AIS, ANSI, ISSIP
	Information management	Karmarkar	MS	BIT	INFORMS
Resource organizations	Economic geography	Bryson	SIJ		
	Service systems	Ng	EMJ	Alliance	
	Social enterprises	Lyons	HFEMSI	HSSE	AIS, ISSIP

Journals: *CACM* Communications of the ACM, *Computer* IEEE Computer, *ECRA* Electronic Commerce Research and Applications, *EMJ* European Management Journal, *HFEMSI* Human Factors and Ergonomics in Manufacturing & Service Industries, *IBMSysJ* IBM Systems Journal, *IJIMA* International Journal of Internet Marketing and Advertising, *IJSIM* International Journal of Service Industry Management, *IJISSS* International Journal of Information Systems in the Service Sector, *ISEBM* Information Systems and E-Business Management, *MS* Management Science, *JAMS* Journal of the Academy of Marketing Sciences, *JBIM* Journal of Business & Industrial Marketing, *JOSM* Journal of Service Management, *JSR* Journal of Service Science, *JSSE* Journal of Systems Science and Systems Engineering, *MSQ* Managing Service Quality, *OMJ* Operations Management Research, *SIJ* The Service Industries Journal

Conferences: *Alliance* Cambridge Alliance Conference, *AHFE* Applied Human Factors and Ergonomics Conference, *AMA SERVIG* AMA SERVIG Conference, *AMCIS* Americas Conference on Information Systems, *ArtSci* Art & Science of Service Conference, *Frontiers* Frontiers in Service Conference, *HICSS* Hawaii International Conference for Systems Sciences, *HSSE* AHFE Human-Side of Service Engineering, *Forum* Naples Service Forum, *POMS* Production and Operations Management Society, *QUIS* Quality in Services

Associations: *AAA* American Anthropological Association, *AAAS* American Association for the Advancement of Science, *ACM* Association for Computing Machinery, *AIS* Association of Information Systems, *AMA* American Marketing Association, *ANSI* American National Standards Institute, *ASA* American Statistical Association, *ASVSA* Associazione per la ricerca sui Sistemi Vitali (Viable Systems), *CSS* Cognitive Science Society, *DSI* Decision Science Institute, *EOMA* European Operations Management Association, *FSSL* Finnish Society of Sciences and Letters, *IEE* Institute of Industrial Engineers, *IEEE* Institute of Electrical and Electronic Engineers, *IEEE EMS* IEEE Engineering Management Society, *INFORMS* Institute for Operations Research and the Management Sciences, *ISQA* International Service Quality Association, *ISSIP* International Society of Service Innovation Professionals, *NAE* US National Academy of Engineering, *NAPA AAA* National Association for the Practice of Anthropology, *NYAS* New York Academy of Sciences, *OASIS* Advancing Open Standards for the Information Society, *SSEBA* Swedish School of Economics and Business Administration

References

Adams J (2000) Risk. Routledge, London

Angier N (1998) When nature discovers the same design over and over, New York Times Science section. Published: December 15, 1998. http://www.nytimes.com/1998/12/15/science/when-nature-discovers the same design over-and-over.html

Arthur WB (2009) The nature of technology: what it is and how it evolves. Free Press, New York

Auerswald PE (2012) The coming prosperity: how entrepreneurs are transforming the global economy. Oxford University Press, Oxford

Bardhan I, Demirkan H, Kannan PK, Kauffman RJ, Sougstad R (2010) An interdisciplinary perspective on IT services management and services science. J Manag Inform Syst 26(4):13–65

Barile S, Polese F (2010) Smart service systems and viable service systems: applying systems theory to service science. Serv Sci 2(1–2):21–40

Boulding KE (1956) General systems theory—the skeleton of science. Manag Sci 2(3):197–208

Box GEP (1979) Robustness in the strategy of scientific model building. No. MRC-TSR-1954, Wisconsin Univ-Madison Mathematics Research Center

Chesbrough H, Spohrer J (2006) A research manifesto for services science. Comm ACM 49(7):35–40

Deacon TW (1997) The symbolic species: the co-evolution of language and brain. WW Norton & Company, New York

Demirkan H, Spohrer JC (2010) Servitized enterprises for distributed collaborative commerce. Int J Serv Sci Manag Eng Tech 1(1):68–81

Demirkan H, Kauffman RJ, Vayghan JA, Fill H-G, Karagiannis D, Maglio PP (2009) Service-oriented technology and management: perspectives on research and practice for the coming decade. Electron Commer Res Appl J 7(4):356–376

Engelbart DC (1995) Toward augmenting the human intellect and boosting our collective IQ. Comm ACM 38(8):30–32

Ferrucci D, Brown E, Chu-Carroll J, Fan J, Gondek D, Kalyanpur AA, Lally A, Murdock JW, Nyberg E, Prager J, Schlaefer N, Welty C (2010) Building Watson: an overview of the DeepQA project. AI Mag 31(3):59–79

Friedman D (2008) Morals and markets: an evolutionary account of modern life. Palgrave MacMillan, New York

Friedman D, McNeill D (2013) Morals and markets: the dangerous balance. Palgrave MacMillan, New York

Gorman M (2010) Trading zones and interactional expertise: creating new kinds of collaboration (Inside Technology). MIT Press: Cambridge, MA

Hakansson NH, Kunkel JG, Ohlson JA (1982) Sufficient and necessary conditions for information to have social value in pure exchange. J Finance 37(5):1169–1181

Hawley AH (1986) Human ecology: a theoretical essay. University of Chicago Press, Chicago

Hirshleifer J (1971) The private and social value of information and the reward to inventive activity. Am Econ Rev 61:561–574

Hutchins E (1995) Cognition in the wild. MIT, Cambridge

IBM (2011) The invention of service science. http://www-03.ibm.com/ibm/history/ibm100/us/en/icons/servicescience/

Jones BF (2005) The burden of knowledge and the 'death of the renaissance man': is innovation getting harder? NBER Working Paper, No. 11360. May

Kremer M (1993) Population growth and technological change: one million BC to 1990. Q J Econ 108(3):681–716

Maglio PP, Spohrer JC (2013) A service science perspective on business model innovation. Industrial Market Manag 42(5):665–670

Maglio PP, Srinivasan S, Kreulen J, Spohrer J (2006) Service systems, service scientists, SSME, and innovation. Comm ACM 49(7):81–85

Maglio PP, Vargo SL, Caswell N, Spohrer J (2009) The service system is the basic abstraction of service science. Inform Syst E Bus Manag 7(4):395–406

Magnini VP, Ford JB, Markowski EP, Honeycutt ED Jr (2007) The service recovery paradox: justifiable theory or smoldering myth? J Serv Market 21(3):213–225

March JG, Simon HA (1958) Organizations. Wiley, New York

Motwani J, Ptacek R, Fleming R (2012) Lean sigma methods and tools for service organizations: the story of a cruise line transformation. Business Expert Press, Burlington, VT

Mulgan G (2010) Measuring social value. Stanford Soc Innov Rev 8(3):38–43. http://www.ssireview.org/articles/entry/measuring_social_value

Newell A, Simon HA (1976) Computer science as empirical inquiry: symbols and search. Comm ACM 19(3):113–126

Ng ICL (2012) Value and worth: creating new markets in the digital economy. Innovorsa Press, Warwick, UK

Ordanini A, Parasuraman P (2011) Service innovation viewed through a service-dominant logic lens: a conceptual framework and empirical analysis. J Serv Res 14(1):3–23

Ostrom E (2009) Understanding institutional diversity. Princeton University Press, Princeton

Ostrom AL, Bitner MJ, Brown S, Burkhard K, Goul M, Smith-Daniels V, Demirkan H, Rabinovich E (2010) Moving forward and making a difference: research priorities for the science of service. J Serv Res 13(1):4–36

Ricketts JA (2012) Reaching the goal: how managers improve a services business using Goldratt's theory of constraints. Pearson Education, IBM Press, Upper Saddle River, NJ

Rust RT, Zeithaml VA, Lemon KN (2000) Driving customer equity: how customer lifetime value is reshaping corporate strategy. Free Press, New York

Simon HA (1996) The Sciences of the Artificial. MIT, Cambridge

Spohrer J (2009) Service science and systems science: perspectives on social value. Proceedings of the 5th symposium of the 21st century COE program "Creation of agent-based social systems sciences" on February 27 and 28 2009, Tokyo Institute of Technology, Tokyo, pp 9–33

Spohrer J (2010) Whole service. http://service-science.info/archives/1056

Spohrer J (2012) A new engineering-challenge discipline: rapidly rebuilding societal infrastructure. http://service-science.info/archives/2189

Spohrer JJ, Demirkan H (2013) Understanding value co-creations and service innovations in time & space complexity: the abstract-entity-interaction-outcome-universals (AEIOU) theory. Unpublished Working Paper Available on Request

Spohrer JC, Giuiusa A (2012) Exploring the future of cities and universities: a tentative first step. In: Proceedings of workshop on social design: contribution of engineering to social resilience, May 12. System Innovation. University of Tokyo, Tokyo

Spohrer JC, Kwan SK (2009) Service Science, Management, Engineering, and Design (SSMED): an emerging discipline—outline & references. Int J Inform Syst Serv Sector (IJISSS) 1(3):1–31

Spohrer JC, Maglio PP (2010) Toward a science of service systems. In: Maglio PP, Spohrer JC (eds) Handbook of service science. Springer, New York, pp 157–194

Spohrer JC, Maglio PP, Bailey J, Gruhl D (2007) Steps toward a science of service systems. IEEE Comput 40(1):71–77

Spohrer JC, Anderson L, Pass N, Ager T (2009) Service science and S-D logic. Proceedings of the 2009 Naples forum on service, June 16–19 2009, Italy

Spohrer JC, Demirkan H, Krishna V (2011) Service and Science. In: Spohrer JC, Demirkan H, Krishna V (eds) The science of service systems. Springer, New York, pp 325–358

Spohrer JC, Piciocchi P, Bassano C (2012) Three frameworks for service research: exploring multilevel governance in nested, networked systems. Serv Sci 4(2):147–160, June 2012

Tang V (2012) Survey: service science top open questions. http://service-science.info/archives/2071

Tracy S (2011) Service systems & social enterprise (MI thesis). University of Toronto. https://tspace.library.utoronto.ca/handle/1807/31608

Tracy S, Lyons K (2013) Service systems and the social enterprise. Hum Factors Ergon Manuf 23:28–36. doi:10.1002/hfm.20516

Van Lange PAM (1999) The pursuit of joint outcomes and equality in outcomes: an integrative model of social value orientation. J Pers Soc Psychol 77:337–349

Vargo SL, Lusch RF (2004) Evolving to a new dominant logic for marketing. J Market 68(1):1–17

Vargo SL, Lusch RF (2008) Service-dominant logic: continuing the evolution. J Acad Market Sci 36(1):1–10

Whitehead AN (1911) An introduction to mathematics. Oxford UK: Oxford University Pres

Wright R (2000) Non-Zero: the logic of human destiny. Vintage/Random House, New York

Chapter 2
Translational and Trans-disciplinary Approach to Service Systems

Kyoichi Kijima

Abstract The aim of this chapter is to examine a research scheme on service systems, or "service systems science," from a translational systems science standpoint. The innovative service science (or service science, management, and engineering: SSME), which treats services as a system and discusses such systems from the point of view of "service-dominant logic," has formed the basis of research into services since its proposal and has expanded significantly from the service marketing field. On the other hand, systems science treats its subjects as systems and not only examines the properties of such systems using interdisciplinary approaches from a holistic point of view, but rapid developments are being made in "translational systems sciences," which connect processes from logic and concepts to theory and modeling, all the way through actual practice and implementation. Within this, service systems science is a service science that emphasizes the framework of such translational systems sciences.

In this chapter, we first summarize claims regarding services in various academic circles. Then we examine these in relation to service science. Next we explain the approach of "translational systems sciences" that forms the backbone of service systems science. We subsequently examine in detail the objectives, research maps, research domains, etc., of service systems science, in comparison with service science. Finally, we illustrate as specific systemic reference models for service systems "Four-phase Value Co-creation Process Model" and "Value Orchestration Platform Model" and argue their significance.

Keywords Service systems science • Value co-creation • Translational • Social value • Value orchestration platform

1 Introduction

1.1 Service Science as SSME

Service is increasingly important to many fields. However, each specialization sees service somewhat differently (Spohrer 2009). Economics and the social sciences have distinguished service from agriculture and manufacturing, and service is often

K. Kijima (✉)
Tokyo Institute of Technology, Tokyo, Japan
e-mail: kijima@valdes.titech.ac.jp

© Springer Japan 2015
K. Kijima (ed.), *Service Systems Science*, Translational Systems Sciences 2,
DOI 10.1007/978-4-431-54267-4_2

Table 2.1 Products and services (Grönroos 2000)

Products	Services
Tangible	Intangible
Homogeneous	Heterogeneous
Production and distribution separated from consumption	Production, distribution, and consumption simultaneously
Nonperishable, can be kept stock	Perishable, cannot be kept stock
A thing	An activity or process
Core value produced at a factory	Core value produced in provider-customer interaction
Customers do not participate in production	Customers participate in production
Transfer of ownership	No transfer of ownership

captured as the residue of agriculture and manufacturing actions (Spohrer 2009). Measurements of the growth of the service sector are made in terms of the numbers and types of jobs (employment) and firms (sector growth and competitiveness), contributions to the GDP, and the balance of trade. A well-known comparison of certain aspects of the difference between products and services in economics is illustrating a dichotomy between the two (Table 2.1) (Grönroos 2000, 2007).

Industrial engineering, management science, and operations research emphasize mathematical modeling of and engineering approach to service systems. For example, networks of stochastic service systems are intensively modeled on the basis of their capacity and demand characteristics, while the queuing theory is often adapted to analyze the service capacity. Computer scientists and information systems specialists in particular focus on web services and service-oriented architectures (SOA).

In the psychological and behavioral sciences, service is seen in the context of customer-provider interactions as something to be experienced, remembered, and evaluated. The experience of customer-provider interactions is being increasingly designed.

In service marketing and operations management, a well-known conceptual model, namely, the Gap Model of Service Quality, is used to understand the service quality of an organization (Parasuraman et al. 1984). The model identifies a customer gap and four provider gaps and then claims that these five gaps must be closed in order to increase the service quality. According to the model, closing the customer gap, which is defined as the difference between the customer's expectation and the actual experience of the service, is the most critical to delivering quality service.

As briefly reviewed so far, the concepts of, interests in, and approaches to service are quite diversified among disciplines. Service science, or more precisely, service science, management, and engineering (SSME), is an emerging area of study that draws on pioneering works in service marketing, service operations, service management, service engineering, service economics, and service computing (Maglio and Spohrer 2008; Spohrer and Kwan 2009). Service science tries to shed light on a scientific approach to understanding social value and identifying propositions that can be formulated and theories that can be empirically tested. Service science defines service observable in the world in terms of a service system with value co-creation interactions among entities by taking a bird's-eye view of various perspectives. A service system is a dynamic interaction between such providers and customers as people, businesses, nonprofits, government agencies, and even cities based on shared information with support of ICT (IfM and IBM 2008). Value and value co-creation are at the heart of service and are critical for understanding the dynamics of service systems and furthering service science.

The goal of service science is to promote service innovation by interdisciplinary approach. Innovation is a key to productivity, quality, and contents of service and emerges from the intersection of different types of knowledge. To this end, promoting an interdisciplinary approach is crucial to the field.

1.2   Service in Service Science

A basic stance of service science to service itself can be expressed by service-dominant (S-D) logic (Vargo and Akaka 2009; Vargo et al. 2008; Lusch et al. 2010). According to the logic, service is defined as an experience and a phenomenon. It claims, "Value is always uniquely and phenomenologically determined by the beneficiary." The roles of producers and consumers are not distinct, because value is always co-created—jointly and reciprocally—in interactions among providers and beneficiaries through the integration of resources and application of competences.

Individual service experiences are embedded in specific individual and social contexts. Indeed, service experiences are both intra- and intersubjective because individuals do not live in isolation but rather as part of different groups and networks.

Another streams of service science, Nordic School Approach and New Service Development (NSD), understand service as a process. Value co-creation is an active, creative, and social process based on collaboration between the provider and customer that is initiated by the provider to generate value for customers. Such collaborative value co-creation often requires greater effort on the part of both the customer and provider than does a traditional market interaction. The people on both sides must think about what they want to get out of a cooperative relationship. The perspective of service as a process helps managers advance the service co-creation and allocate resources, as well as decode and record architectural elements and phases of innovation (Helkkula et al. 2012).

These two perspectives about service are not exclusive and are rather used to provide different, rich, and complementary angles to understanding service (refer to Table 2.2) (Helkkula et al. 2012). Table 2.2 also includes goods-dominant logic to compare S-D logic and Nordic School Approach and New Service Development to it.

Table 2.2 Goods-dominant logic, S-D logic, Nordic School Approach, and New Service Development

Perspective/paradigm/logic	Goods-dominant logic	Service-dominant logic	Nordic School Approach and New Service Development (NSD)
What is service?	Service as an outcome (new kinds of service products or attributes)	Service as an experience (valuable, subjective experiences in different events)	Service as a process (a new, well-functioning process)
Emphasis	Service is measured by attributes and variables in a functional domain	Value is always co-created, jointly and reciprocally in interactions	Value co-creation is an active, creative, and social process based on collaboration
In what is value created?	Value in exchange	Value in context	Value in use
		Value in experience	

2 Translational Systems Sciences

Translational systems science is a new trend within systems sciences motivated by the need for practical applications that help people by a holistic, comprehensive, and systems thinking on problematic complexity.

The concept of translational research originally comes from medical science for enhancing human health and well-being. In the field it is used to translate the findings in basic research more quickly and efficiently into medical practice and meaningful health outcomes. Translational medical research in this direction is often labeled as "bench to bedside" since it involves the movement of research from laboratory to the clinical practice, i.e., theories emerging from preclinical experiments are tested on disease-affected human subjects. Translational medical research is, however, a bidirectional process, that is, the first one goes from "bench to bedside" and then from "bedside to bench," where information obtained from human experiments is used to refine the understanding of biological principles, heterogeneity of human disease, and polymorphism.

Core of translational systems sciences comes from much broader translational systems thinking based on the platform underpinned by three domains, i.e., systems concepts/models/theories, systems methodologies, and systems practice (Fig. 2.1).

They correspond to the way in which thinking in general has been philosophically framed, i.e., episteme (know why), techne (know how), and phronesis (know when, know where, know whom) (Ing 2012) (See Table 2.3).

Fig. 2.1 Three dimensions of translational approach

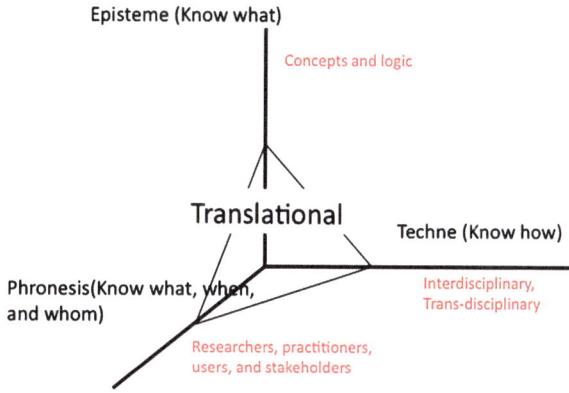

Table 2.3 Episteme, techne, and phronesis as primary intellectual virtues

Thinking in general			
Primary intellectual virtue	Episteme	Techne	Phronesis
Translation/interpretation	Science (viz., epistemology)	Craft (viz., technique)	Prudence, common sense
Type of virtue	Analytic scientific knowledge	Technical knowledge	Practical ethics
Orientation	Research	Production	Action
Nature	Universal	Pragmatic	Pragmatic
	Invariable (in time and space)	Variable (in time and space)	Variable (in time and space)
	Context-independent	Context-dependent	Context-dependent
Pursuits	Uncovering universal truths	Instrumental rationality toward a conscious goal	Values in practice based on judgment and experience
Colloquial description	Know why	Know how	Know when, know where, know whom
Systems thinking			
Categories of systems thinking	Systems models/theories (e.g., living systems theory, hierarchy theory, open systems theory)	Systems methodologies (e.g., hard systems approach, soft systems approach)	Systems practice (e.g., action research, structured dialogic design)

Hence, the term "translational" implies that translational systems science is scientific research that facilitates the translation among concepts/models/theories, methodologies, and practices. With its focus on removing barriers to interdisciplinary methodologies, translational research has the potential to drive the advancement of applied as well as basic science.

3 Service Systems Science

Service systems science is a study of social value co-creation phenomena/process among service system entities in translational systems science perspective. Social value here is a broad concept that includes not only economic values such as revenues and profits but also psychological fundamental values such as safety and security, cultural and emotional value such as empathy, and even such social values as sustainability relevant to all mankind (Spohrer and Maglio 2008).

As a typical research area of translational systems sciences, service systems sciences attempt to approach to social value co-creation phenomena/process by drawing on, bridging, and integrating (1) systems theories and models, (2) systems methodologies, and (3) systems practices in such a holistic way that it tries to derive understanding of parts from the behavior and properties of wholes rather than derive the behavior and properties of wholes from those of their parts.

3.1 Service Systems Theories and Models

Now we will argue the meaning of identify service as a system, or of applying translational systems science perspective to service systems.

Although differences exist between the short and long term, classical economic and business research on services has focused on monetary values. However, rather than just limiting "service" to the narrow sense of the word as "the service industry," by classifying it as creative activities with a broader range of values, it is possible to treat these various values, such as the so-called psychological and social ones, in a multifaceted manner. By so doing, "service" is able to be assessed according to multiple values, not just monetary and economic measures; and it becomes possible to incorporate the concept and methodologies of systems science. For example, soft systems approach such as SSM (soft systems methodology) can be put to practical use in the design of social services used by stakeholders who have a multitude of different values (Checkland 1999; Jackson 2003). To put it more fundamentally, systems methodologies of thought can be incorporated into hands-on activities such as innovation and service improvement.

By focusing on "systems characteristics," we can provide common ground for the discussion of a wide range of services. Treating services as "systems" means focusing on the nature of subjects, such as those described below, as systems characteristics, and observing these subjects. One characteristic of systems is that they provide a common language when describing a variety of services. This can provide a common and commensurable ground and perspective when a wide range of stakeholders are in discussion.

Bertalanffy categorized systems as *closed*, i.e., systems having no interaction with other (external) environments, and *open*, i.e., systems that interact with external environments, including other systems, in the process adapting themselves to their environment (Bertalanffy and Sutherland 1974; Bertalanffy 1968).

The concept of open systems also exerts considerable influence on the study of management organizations. For example, "3C" and "SWOT Analysis," which form the most basic framework in management studies, are tools to help us understand relationships between organizations and their environments.

The question of whether we treat service systems as open systems, or closed system, runs parallel to the issue of where and how we should set the *boundaries* of the service system concerned. The concept of such boundaries is fundamental to how systems sciences perceive things. If boundary conditions change, then relationships change, too. For example, in a decision-making problem, the order in which we choose alternative proposals will change depending on how boundaries are drawn. Alternatives that maximize an individual's utility are not necessarily equivalent to alternatives that maximize social well-being.

Systems sciences take an epistemological view toward *complexity* and *hierarchy*; complexity is not assumed a characteristic inherent in the subject; rather it is determined in interactions with observers (Klir 2001). Klir states that "Complexity of a system exists in the eyes of the observer," as explained in the following example. When a neurophysiologist looks at the brain of a sheep, he/she sees an extremely complex system with the interaction of the neural networks and vast numbers of neurons, yet when a butcher looks at it, at best he/she sees a type of meat to be divided into dozens of cuts.

In service systems, the fact that the service providers and consumers view complexity in different ways can easily cause problems.

Such complexity is closely related to the hierarchy of the system. If we regard a certain single system as a component, we can think of it as a high end system (stand-alone, superior, upper class, ranked), or we can make the system nested and hierarchical. The deeper the hierarchy is, the greater the complexity is certainly perceived.

The hierarchical level at which we view a system is very important and must be determined appropriately. For example, in considering service systems, it is pointless to view them on a cell level.

Communications and control itself is an old concept, but Wiener advocated cybernetics as a communication and control theory dealing with various systems (Wiener 1965). One of the essential core concepts is that of positive and negative feedback. Further, the Law of Requisite Variety proposed by Ross Ashby reached a significant theoretical point in the field known as organizational cybernetics (Ashby 1958). This claims that in order to counter a system's external diversity, there must be sufficient internal diversity provided. In debates on service systems, the claim that it is important to diversify by means of flexibility in service systems in order to meet the diverse needs of customers is supported by this Law of Requisite Variety.

In the context of the Law of Requisite Variety, the concept of the internal model is an important one (Kijima 1986). An internal model is one constructed by the subject in relation to the environment (usually subjectively) and is also known as an interpretation model or mental model. To state the Law of Requisite Variety differently, control only becomes possible when a control system inherently encapsulates some representation (recognition) of the process to be controlled (known as the Principle of Internal Modeling (Wonham 1963; Francis and Wonham 1976)).

Service providers and their customers can share an internal model, by making essential mutual understanding of what kind of service will be supplied and what kind of service is required. With this mutual understanding established, both parties can then begin to co-create value for both parties (this point is also clearly expressed by the Four-phase Value Co-creation Process Model to be mentioned later).

In this way, when the systems properties, or, more broadly, the systems concepts, comprehend and recognize a large number of service systems, it provides an effective common language. In fact, when a variety of professionals with expertise in different fields collaborate with each other, common language provided by systems sciences and systems properties should become effective communication language to specify problems and solve challenges.

3.2 Service Systems Methodologies and Practice

Methodologies for systems practices have been broadly divided into "hard systems approach," which primarily uses quantitative systems models and optimizes them assuming what problems are already obvious, and "soft systems approach," which, encouraging stakeholder participation, primarily uses qualitative systems models to structure problems where the issues are not obvious (Checkland 2000; Rosenhead 2000).

In the study of service systems science, both the hard systems approach and the soft systems approach are necessary. In particular much of the research currently taking place in Japan on so-called service engineering visualizes the service situation and introduces quantitative management like that in the manufacturing industry, adopting a similar position to the hard systems approach. Fähnrich et al. define service engineering research topics as "optimization existing service efficiency," "measurement and modeling of human behavior and values," and "clarification of service/innovation/mechanism and value creation" and show optimization trends as represented in operations research (Fähnrich and Meiren 2007).

Examples of problem solving based on this type of service engineering include an IC chip placed in dishes that rotate on sushi bar conveyor belts, which improves product quality by removing dishes that have been rotating on the belt for some time since their preparation, as well as using demand forecasts based on customer demographics to reduce waste. However, for service systems whose main service resources include human beings, optimization methods are not necessarily appropriate. In particular, many studies attempt to optimize the "obvious" KPIs (key performance index) without identifying exactly what should be optimized.

In contrast, the soft systems approach is concerned with what the relevant issues are and what should be the problems. In an older but well-known example, Ackoff argues the problem of elevator waiting times in a condominium (Ackoff 1981). Residents of a certain apartment block became increasingly dissatisfied with the length of time they must wait for an elevator. However, the solution was not to install an additional elevator, nor was it the optimization of elevator movement as optimization analysis suggested. In the end, the problem was solved by installing a mirror in the elevator lobby. In other words, in this story the issue was not the shortening of waiting times per se; the issue was emotional in the sense that people had nothing to do while they waited.

Many specific methods have been proposed for the soft systems approach. The VSM (viable system model) (Beer 1985) and SSM (soft systems methodology) (Checkland 1999; Jackson 2003) are well-known representatives. In terms of improvements to service systems, there are many examples of SSM being applied to government services. Moreover, in recent years VSM (Beer 1994) has enjoyed popularity as a means of diagnosing and designing service systems.

3.3 Service Systems Science and Service Science

Service science is a specialization of systems science in the sense that the former restricts its attention to artificial human-made worlds (Spohrer 2009; Spohrer et al. 2008). Service systems science is service science explicitly approached from translational systems science perspective. In comparison with service science and its underlying service-dominant logic, what are the specific characteristics and points of emphasis in service systems science?

Firstly, it explicitly tries to describe and understand service in terms of systemic properties illustrated in Sect. 3.1. It claims, among others, that if service is identified and formulated as service "system," then we can conduct deeper research on it adopting rich knowledge developed by systems sciences so far. The "systemic way of looking at services" is fundamentally emphasized.

Secondly, compared to service science, service systems science emphasizes the point that the values incorporated therein are social values. Economic activity is often discussed in service science. In service systems science, social activity is not viewed one dimensionally in terms of monetary value; rather it takes the stance that social activity has multiple attributes and is multidimensional.

Another significant characteristic of service systems science is its translational approach. As we have already seen, service systems science has been developed within a framework of translational systems science that incorporates the engineering and practice phases as well as modeling and theorizing phase. The "systemic way of looking at things" has a cyclical structure to research that leads from theory and models (episteme) to the realization of service innovation (techne) and practical applications via action research methodology on systems intervention (phronesis).

Finally, the concept of the internal model is highlighted in service systems science. A so-called desirable environment, in terms of a social principle, may not necessarily be "desirable" in terms of other principles or when viewed from upper/lower echelons of the hierarchy. With regard to the "desirability" embraced in this principle, if various other principles have their own respective internal models, it is possible for recognition errors to occur. Such subjective internal models as well as reciprocal learning are clearly expressed in the four-phase process model of value co-creation as described below.

3.4 Service Systems Science: Research Map and Domains

3.4.1 Research Map of Service Systems Science

Table 2.4 shows a specific research map of service systems science, from episteme to techne to phronesis. Category 1 on the map is the domain dealing with episteme and techne. Here, the common concepts, theory, and models of service systems are examined, and depending on the complexity of the subject, a wide variety of methods are developed such as modeling methods that use a mathematical approach, simulation, and conceptual modeling.

On the other hand, Category 2, rather than dealing with episteme and techne, is a more practical and implementation domain concerned with research into the design and improvement of service systems, taking into account the unique characteristics of any given situation and depending on the service system context. Here, (i) the hard systems approach which aims for optimization and efficiencies based on problem-solving paradigms such as service engineering and (ii) the soft systems approach which tries to clarify the whereabouts of a problem via the mutual understanding of the various stakeholders involved are used in a complementary way.

3.4.2 Research Levels and Domains of Service Systems Science

In service systems science, three research levels are defined in accordance with how service entities and interactions are treated, based on the hierarchy of the service system.

Micro Level

Micro level service systems are those whose factors comprise a provider who proposes the co-created value and a consumer who agrees to that proposal. Its objective is the analysis of the interaction between the provider and the consumer. Providers

Table 2.4 Research map for service systems science

		Examples
1. Study of service systems as such in systems theory perspective (episteme and techne)	1.1 Service philosophy, concepts, logic, or paradigm	• Service-dominant logic
	1.2 Service theories and models	• ISPAR (Interact-Service-Propose-Agree-Realize) model (Spohrer and Kwan 2009)
	1.2.1 Mathematical models	• PCN (Process Chain Network) model (Sampson 2010)
	1.2.2 Simulation models	• Four-phase value co-creation process model
	1.2.3 Conceptual models	• Value orchestration platform model
2. Applications of service systems ideas with systems methodologies and systems practice (techne and phronesis)	2.1 Systemic (soft) approach to problem structuring of service system	• Service collaboration • Scenario writing
	2.2 Systematic (hard) approach to problem solving of service system	• Service engineering • Service efficiency measurement

and consumers refer strictly to roles, and the subject of analysis is not limited to individual persons. The main point of debate under this model is the decision-making between the provider and the consumer, and the issues are "under what conditions does service interaction occur?" and "is such interaction sustainable in the long term?"

Social value co-creation phenomenon/process at the micro level service systems includes at least the following research domains: The first domain is related to such basic value of social infrastructure as security, safety, amenity, life, and health. The value is fundamentally necessary for every human being to enjoy everyday life happily and is provided, for example, by electric power systems, water and sewer services, transportation systems, education, health care, and finance information systems.

The second domain focuses on innovative value of business models for private companies, nonprofit organizations, and public sectors to promote free and fair social and economic activities by realizing smart society, smart city, and smart government.

The third domain is concerned with the value of global community, which deals with consensus building and confrontation management for solving such global issues as the environmental, energy, and/or food problems to realize happy and sustainable global community (Table 2.5).

Table 2.5 Research domains at micro level

Research domains	Examples of research topics
Basic value of social infrastructure	Terrorism and risk management, health care, local revitalization, education, utility systems
Innovative service value of business	Innovative business models, service at NPO, service marketing
Sustainable development of global community	Consensus building, confrontation management, global environment

Meso Levels

Meso level analysis uses as a component factor the "service value" arising from interactions at the micro level and analyzes the related integration and decomposition factors. Important issues at this level include integration methods for interactions at the micro level and problems of flexibility and the handling of diversity arising from a change in combinations. Moreover, an important aspect is the analysis of the flow of costs and added value that occur when interactions are integrated.

Macro Levels

Further, the approach at the macro level is to take the indices from the results of the meso level and use the aggregated service statistics for a certain group and over a certain time period, to analyze the correlations between indices. A typical example of the macro level is one whereby a nationwide service in Japan is suffering from sluggish labor productivity and statistical models are used to analyze the contributory factors. The purpose of debate at the macro level is to use such analyses in policy making.

Many of the issues faced by service systems are connected to multiple hierarchical layers. For example, if there is a debate on the quality and level of satisfaction with regard to a service supplied by a company, the decision-making process is analyzed in a micro model on the interaction between the provider and the consumer using questionnaires, and this clarifies the mechanisms. At the meso level, by changing the combination of the interaction mechanism, we can analyze how to change such satisfaction levels and lead ourselves to new services and designs. The macro level is implemented via comparative studies done at the national level (by the JCSI (Japanese Customer Satisfaction Index) in Japan) that look at these satisfaction levels within the same industry and between different types of industry. The level of analysis we choose is very important in getting our desired results and outcome from the analysis.

4   Four-Phase Value Co-creation Process Model

Four-phase Value Co-creation Process Model is a model for opening up the concept of value co-creation (Galbrun and Kijima 2009a, b). We identify value co-creation interaction as an active, creative, and social process based on the collaboration between the provider and customer that is initiated by the provider to generate value for customers. It is a form of collaborative creativity of customers and providers that enhances knowledge-acquisition processes by involving the customer in the creation of meaning and value, although it is initiated by the provider.

Such collaborative value co-creation often requires greater efforts on the part of both customer and provider than does a traditional market interaction. People on both sides must think about what they want to get out of a cooperative relationship. Customers need to trust the provider to not misuse the information they provide or unfairly exploit the relationship. Since providers need to actively manage customer expectations about how the relationship will evolve, they must be so trained as to have capabilities for efficiently co-creation.

However, it may be too simple to assume that both sides know about the other's preference, expectations, or capabilities when participating in the collaborative process. Rather, they may or may not need to learn about each other to share internal models (mental models).

This consideration leads us to the idea of service as a dynamic interaction process in which customers and providers are mutually learning and collaborating by co-experience.

Now, we propose a new model called the "Four-phase Value Co-creation Process Model" (refer to Fig. 2.2).

The model explicitly defines service as a value co-creation interaction between customers and providers and identifies four phases that occur in the process. The first two phases, co-experience and co-definition, are relatively short-range concepts for describing service appreciation, while the final two phases, co-elevation and codevelopment, refer to the long-range activities necessary for service innovation.

Fig. 2.2 Four-phase value co-creation process model

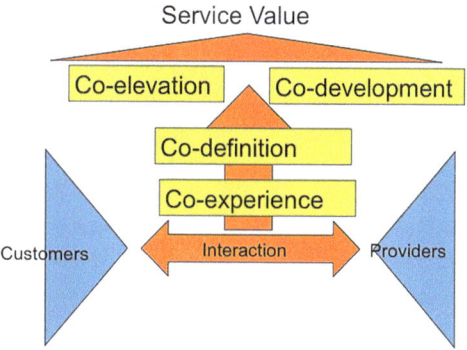

4.1 Co-experience of Service

When participating in the collaborative value co-creation process, customers and providers may have little or no idea about the other's capabilities and expectations. Hence, rather than reducing the gap between the needs and seeds, by co-experience, the provider and customer share an internal model to co-define a mutual understanding about the service.

4.2 Co-definition of Shared Internal Model

By interacting with each other, the customer and provider may learn about the other's preference, capabilities, and expectations so that they may co-define and share a common internal model (Chesbrough and Spohrer 2006; Vargo and Akaka 2009).

Satisfaction for both sides is generated by the co-experience of the service and the co-definition of shared internal model. For example, at a sushi bar, through conversation, the chef recognizes a customer's preferences, mental and physical condition, and appetite and the customer learns about the day's specialties and seasonal fish. If they are able to share a common internal model (i.e., understand the other's preferences, capabilities, and expectations), then both are satisfied. This is a typical process of co-experience and co-definition.

4.3 Co-elevation of Each Other

Co-elevation is a zigzag-shaped spiral process of customer's expectations and provider's capabilities. Higher expectations of service by the customer lead to higher-quality service and greater social values (needs pull). High-quality service, in turn, increases customer expectations level (seeds push). For example, in the mornings, Tokyo commuter subway trains arrive and depart every 3 min and 40 s. The driving force for such punctual, safe, and frequent service is the high level of service demand by customers and the provider's ability to meet such requirements.

4.4 Codevelopment of Value

On the other hand, we call the latter codevelopment because it pays attention to the value co-created by simultaneous collaboration among various entities. Codevelopment of service value is usually carried out in the context of customers evaluating and assessing the value and providers learning from customer responses. Collaborative improvement of Linux software or Wikipedia by anonymous engineers and developers is a typical example of codevelopment.

Fig. 2.3 Value orchestration
platform model

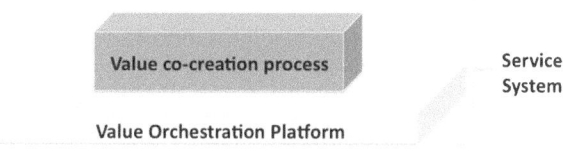

5 Value Orchestration Platform Model for Value Co-creation Management

Value Orchestration Platform Model is a model for analyzing new service businesses, whether real or virtual, based on the establishment of sophisticated logistics and the rise of information technology (Kijima et al. 2012, 2013).

One common characteristic of the new service businesses is that they have two layers (see Fig. 2.3). As shown in the figure, in the value co-creation process at the top layer, customers and providers interact with each other and co-create new values. The process is opened up in the previous section.

The bottom layer invites customers and providers to "get on board."

It facilitates and orchestrates new value co-creation by customers and providers, but leaves the control of the process entirely in the hands of providers and sometimes of customers as well. We call this layer the value orchestration platform.

Though the essential idea of a value orchestration platform dates back several decades and its traditional and well-known examples include credit cards and shopping malls, we may see that websites such as Amazon, eBay, iTunes, Rakuten, and YouTube primarily are typical examples of value orchestration platforms. They connect tens of thousands of providers (sellers) to millions of customers (buyers). For example, the strength of the Apple App store lies in its function as a value orchestration platform. The store is interested in inviting as many users and appropriate developers as possible onto it, but it takes no physical or full legal "possession" of the software it distributes.

In the following part of this section, we discuss value orchestration management strategies (i.e., involvement, curation, and empowerment) in terms of service systems science perspective, by referring to actual cases in which such strategies are implemented.

5.1 Involvement Strategy

Platform orchestrator is primarily concerned with the methods to get appropriate customers and providers "on board" the platform and to vitalize interactions between customers and providers. Hence, strategies for the platform to attract and involve customers and providers to maximize profit are crucial. Indeed, one of the advantages of an online value orchestration business such as e-commerce is that they have no limitations on the number of customers who can participate.

A cycle of sympathize, identify, participate, share, and spread (SIPS) is useful for identifying how customers and providers become interested in a platform. It generates interest among customers and the providers toward co-experience and co-definition phases (Dentsu Inc. 2011).

SIPS proposes that the trigger for customers and providers to become interested in a service system is their having sympathy toward it. Presently, people are connected with each other through social media outlets such as Facebook and Twitter. They communicate through rather subjective comments about what they experience, and the comments that gain a certain level of sympathy for being useful and interesting spread quickly throughout these media platforms. As a result, the media triggers sympathy to, for example, a shopping mall and leads to its identification as an interesting place.

In SIPS, participation does not necessarily mean purchase of some products or services. Rather, it emphasizes that the experience would lead to sharing and spreading through a common internal model.

5.2 Value Curation Strategy

While strategy for the involvement of customers and providers focuses on how to attract customers and providers on the platform, value curation is essential for the platform to encourage customers and providers to co-elevate and codevelop.

Curation can be defined as a highly proactive and selective approach of value orchestration that collects, selects, analyzes, edits, and reexamines the content and meaning of existing products, service, and information on customers and providers to provide a new interpretation of and a new meaning to them. Based on the newly developed interpretation and meaning, it facilitates a value co-creation process involving customers, providers, information, and technology.

To collect information, sufficient technology and methodology is necessary for scooping up appropriate information from an enormous amount of data on the Internet and databases. To provide a new interpretation of the information, it is necessary to combine human intelligence with technology to evaluate, understand, and process data; dig out information and value from that data; and visualize what the data indicates. To facilitate a value co-creation process, multiple approaches to the mental and physical aspects of human beings in both online and offline spaces are essential. Blending new content while filtering and managing other useful information is a productive and manageable solution for providing prospective customers with a steady stream of high-quality and relevant content. While pure creation may be demanding and pure automation does not engage, content curation can provide the best of both.

5.3 *Empowerment Strategies of Stakeholders*

Empowerment is another aspect of value orchestration, particularly for the co-elevation and codevelopment phases. Specifically, this refers to how a platform empowers customers and providers so that each side finds the other attractive and both are motivated to interact with each other. Customers are empowered by lifting up their aspiration level, while providers are empowered by referring to their capability of providing service.

6 Conclusion

This chapter introduced service systems science, a research scheme on service systems based on a translational systems science standpoint. Service systems science is service science that highlights this translational systems science framework. Service science is an ambitious academic undertaking integrating the principles of service-dominant logic, while service systems science adopts a position close to service science but from the perspective of translational systems sciences, which are characterized by a "trans-disciplinary" and "translational" approach.

In this chapter we, in comparison with service science, examined in detail its objectives, research maps, and research domains. Then, we introduced examples of specific systemic models related to service systems as developed by service systems science, namely, the Four-phase Value Co-creation Process Model and Value Orchestration Platform Model, and we considered their significance.

References

Ackoff RL (1981) The art and science of mess management. Interface 11(1):20–26, http://pubsonline.informs.org/doi/abs/1

Ashby WR (1958) Requisite variety and its implications for the control of complex systems. Cybernetica 1(2):83–99

Beer S (1985) Diagnosing the system for organizations. Wiley, Chichester

Beer S (1994) Beyond dispute: the invention of team syntegrity. Wiley, Chichester

Bertalanffy L (1968) General system theory: foundations, development, applications. Braziller, New York

Bertalanffy L, Sutherland J (1974) General systems theory: foundations, developments, applications. IEEE Trans Syst Man Cybern 4(6):592–592

Checkland P (1999) Systems thinking, systems practice. Wiley, Chichester

Checkland P (2000) Soft systems methodology: a thirty year retrospective. Syst Res Behav Sci 17:S11–S58

Chesbrough H, Spohrer J (2006) A research manifesto for services science. Comm ACM 49(7):35–40

Dentsu Inc. (2011) SIPS. http://www.dentsu.co.jp/sips/index.html. Accessed 18 May 2014

Fähnrich K-P, Meiren T (2007) Service engineering: state of the art and future trends. In: Spath D, Fähnrich K-P (eds) Advances in services innovations. Springer, Berlin Heidelberg, pp 3–16

Francis BA, Wonham WM (1976) The internal model principle of control theory. Automatica 12(5):457–465

Galbrun J, Kijima K (2009a) A co-evolutionary perspective in medical technology: clinical innovation systems in Europe and in Japan. Asian J Tech Innovat 17(2):195–216

Galbrun J, Kijima K (2009b) Fostering innovation system of a firm with hierarchy theory: narratives on emergent clinical solutions in healthcare. In: Proceedings of the 52nd annual meeting of the ISSS, Madison

Grönroos C (2000) Service management and marketing. Wiley, West Sussex

Grönroos C (2007) In search of a new logic for marketing. Wiley, West Sussex

Helkkula A, Kelleher C, Pihlström M (2012) Characterizing value as an experience implications for service researchers and managers. J Serv Res 15(1):59–75

IfM and IBM (2008) Succeeding through service innovation: a white paper based on Cambridge service science, Management and engineering symposium and the consultation process. University of Cambridge Institute for Manufacturing (IfM) and International Business Machines Corporation (IBM)

Ing D (2012) Science, systems thinking, and advances in theories, methods and practices. Syst Pract Action Res 12(4):425–430, http://link.springer.com/10.1023/A:1022404515140. Accessed 9 April 2014

Jackson M (2003) Systems thinking: creative holism for managers. Wiley, West Sussex

Kijima K (1986) Algebraic formulation of relationship between a goal seeking systems and its environment. Int J Gen Syst 12:341–358

Kijima K, Rintamki T, Mitronen L (2012) Value orchestration platform: model and strategies. In: Proceedings of human side service engineering 2012, San Francisco

Kijima K, Rintamaki T, Mitronen L (2013) Value orchestration platform and value co-creation process: a hierarchical service systems model and its implications. In: Proceedings of the 2013 forum, Naples Forum on Service, Naples

Kyoichi Kijima, Timo Rintamaki, Lasse Mitronen. (2013) Value Orchestration Platform and Value Co-Creation Process: A Hierarchical Service Systems Model and its Implications, Proceedings of the 2013 Forum, Naples Forum on Service, Naples, Jun. 2013

Klir G (2001) Facets of systems science, 2nd edn. Plenum, New York

Lusch RF, Vargo SL, Tanniru M (2010) Service, value networks and learning. J Acad Market Sci 38(1):19–31

Maglio PP, Spohrer J (2008) Fundamentals of service science. J Acad Market Sci 36(1):18–20

Parasuraman A, Zeithaml VA, Berry LL (1984) A conceptual model of service quality and its implications for future research. J Market 49(4):41–50, http://www.jstor.org/discover/10.2307/1251430?uid=25404&uid=3737976&uid=2&uid=5910312&uid=3&uid=25392&uid=67&uid=62&sid=21103987774637ource=gbs_api

Rosenhead JEA (2000) Rational analysis for a problematic world: revisited. Wiley, Chichester

Sampson SE (2010) The unified service theory. In: Maglio PP, Kieliszewski CA, Spohrer JC (eds) Handbook of service science. Springer

Spohrer J (2009) Service science and systems science. In: Proceedings of COE final symposium, Tokyo, 2009

Spohrer J, Kwan S (2009) Service Science, Management, Engineering, and Design (SSMED): an emerging discipline-outline & references. Int J Inform Syst Serv Sector 1(3): 31

Spohrer J, Maglio PP (2008) The emergence of service science: toward systematic service innovations to accelerate co-creation of value. Prod Oper Manag 17(3):238–246

Spohrer J et al (2008) Service science and service-dominant logic. Otago Forum 2, 8–12 December 2008, University of Otago, Dunedin

Vargo S, Akaka MA (2009) Service-dominant logic as a foundation for service science: clarifications. Serv Sci 1(1):32–41

Vargo SL et al (2008) On value and value co-creation: A service systems and service logic perspective. Eur Manag J 26(3):145–152

Wiener N (1965) Cybernetics or control and communication in the animal and the machine, MIT Press

Wonham WN (1963) Towards an abstract internal model principle. IEEE Trans SMC 6(11): 735–740

Chapter 3
Service Artifacts as Co-creation Boundary Objects in Digital Platforms

Anssi Smedlund and Ville Eloranta

Abstract Digital platforms are systems consisting of a platform owner and complementary and interdependent components. Service artifacts are boundary objects created by the digital platform owner that engage the end user and facilitate the knowledge processes required for value co-creation. These service artifacts function as communication mechanisms in the front end of the virtual platform, operating between the platform and the end user. We present three categories of service artifacts based on their functioning logic and the type of interaction. Working from the theory of boundary objects, we argue that database service artifacts, character artifacts, and artificial intelligence artifacts facilitate personalized value co-creation for each user individually in addition to helping the end user understand the service processes and the variety of offerings available in the platform. We present examples of different types of service artifacts that illustrate these principles. The concept of a service artifact is discussed from the viewpoint of service-dominant (S-D) logic.

Keywords Boundary objects • Digital platforms • Service artifacts • Service-dominant logic

1 Introduction

Platforms are evolving systems that are built around a core technology or service that is essential to broader, interdependent ecosystems of businesses (Gawer and Cusumano 2008). The platform mediates interactions with end users and providers of complementary products or services. The value generated in this interaction is greater than the value that would have been generated through separate interactions, and growth in the number of participants increases the potential of the entire system through the mechanism of network externalities (cf. Shapiro and Varian 1999).

Digital platforms are a type of a platform (Evans 2003). They have become essential for organizing business transactions in a growing number of industries. Betfair.

A. Smedlund (✉) • V. Eloranta
Department of Industrial Engineering and Management,
Aalto University School of Science, Espoo, Finland
e-mail: anssi.smedlund@aalto.fi; ville.eloranta@aalto.fi

© Springer Japan 2015   55
K. Kijima (ed.), *Service Systems Science*, Translational Systems Sciences 2,
DOI 10.1007/978-4-431-54267-4_3

com (Davies et al. 2005) and the Mac App Store are examples of digital platforms that have dramatically changed the industrial structures of their respective businesses. As digital platforms gain more popularity and compete with each other, engaging and collaborating with end users becomes increasingly important to achieving better user experience and loyalty. Platform owners need strategies to lock-in participants and thus secure the evolution and growth of the platform around the core.

Boundary objects exist in the interfaces of organizational borders and help to overcome the inherent imbalances of knowledge between the parties. Boundary objects may be intentionally created, or they can emerge over time during collaboration (Star and Griesemer 1989). Service artifacts are boundary objects created by the platform owner that engage the end user and facilitate the processes of value co-creation in digital platforms. Service artifacts make it easier for the end user to learn from and experience the various possibilities offered by the platform, to develop and refine needs, and to provide feedback on new features for the platform without direct help from personal customer representatives.

An example of a service artifact is the personalized recommendation lists at Amazon.com. This service artifact is a means through which the end user becomes aware of new opportunities for consumption, and because the list evolves based on the customer's browsing and buying history, the artifact evolves as well. Another example is the Moodagent application, which automatically generates music playlists from the 20 million songs stored on the Spotify.com platform based on the current state of mind of the listener. Many Japanese companies have created cartoon characters that communicate with customers and provide a "face" for the company. An example of this type of service artifact is Daimaru Matsuzakaya Department Store's Sakura Panda. The panda is a fictional cartoon character that actively communicates with customers in social media, advertisements, and events. Futuristic service artifacts are artificially intelligent and able to make decisions on behalf of the end user: Apple's Siri is an example of this.

In this paper, we introduce the concept of service artifacts. In Chap. 2, we theorize about the importance of service artifacts in value co-creation in digital platforms, followed by Chap. 3, in which we present how service artifacts are related to other parts of the digital platform structure and how they are related to the service-dominant (S-D) logic (Vargo and Lusch 2004, 2008) paradigm. In Chap. 4, we present examples of existing artifacts and categorizations based on interaction type and complexity. We conclude by addressing how database artifacts, character artifacts, and artificial intelligence artifacts may be used as boundary objects in Chap. 5. In Chap. 6, we present a critical discussion about the concept and suggest future avenues for research.

2 Platforms, Boundary Objects, and Co-creation

Platform participants play distinct roles. End users represent the demand side of the platform (Eisenmann et al. 2008). Platform owners own the core product or service that is the starting point for the evolution of the platform (Ceccagnoli et al. 2012; Gawer and

Henderson 2007). Platform providers such as Internet operators provide access to the platform (Eisenmann 2008). Complementors are also participants; they compose the supply side of the platform (Eisenmann et al. 2008).

Through positive network externalities, growing numbers of end users attract growing numbers of complementors and vice versa, thus increasing the value of the platform for all participants (Shapiro and Varian 1999). Positive network externalities work up to a point in the growth stage of the platform (Liebowitz and Margolis 1994). Aside from attracting participants, platform owners seek to lock them into their platforms. Once the participant is locked in, they incur high surplus costs if they use a similar product or service through a competing platform. These switching costs strengthen the end users' commitment to the platform and prevent them from using several platforms simultaneously.

Vargo and Lusch (e.g., Vargo and Lusch 2004, 2008) argue in their service-dominant (S-D) logic that the appropriate unit of exchange is no longer a tangible good but is, rather, the application of capabilities or specialized human knowledge and skills for the benefit of the recipient. The service business is based on the co-creation of value with customers and other participants (cf. Normann and Ramirez 1993; Prahalad and Ramaswamy 2002; Vargo and Akaka 2009). In the light of S-D logic, co-creation means that the interactions between the firm, its customers, its suppliers, and stakeholders in the market environment combine to create value (cf. Prahalad and Ramaswamy 2004). In S-D logic, service is exchanged for service, and goods indirectly convey service. A service artifact can be thought of as a vehicle to convey service between the platform and the end user.

Co-creation by platform participants is impossible without knowledge processes. In S-D logic, knowledge is an operant resource, a resource that is used to mobilize other resources and an underlying source of value and driver of value creation (Vargo et al. 2010). During business transactions, knowledge is created, transferred, and utilized between individuals and between the units of an organization (e.g., Nonaka and Takeuchi 1995). Based on Grant's (1996) ideas regarding knowledge integration, it can be argued that value co-creation requires the integration of multiple existing knowledge bases to develop an ongoing relationship with the end user. The successful transfer of knowledge from one location to another is central to integration activities (Carlile and Rebentisch 2003).

Prior knowledge affects the successful creation of new knowledge and common understanding (Cohen and Levinthal 1990). Participants who are involved in knowledge processes share common artifacts such as discussion topics, common ideas, organizational routines, and physical objects. These artifacts act as boundary objects (Star 2010; Star and Griesemer 1989). "Boundary objects are objects which are both plastic enough to adapt to local needs and the constraints of the several parties employing them, yet robust enough to maintain a common identity across sites" (Star and Griesemer 1989, p. 393). Boundary objects maintain coherence across intersecting worlds. Boundary objects are powerful tools for identifying collaborative settings among participants with different levels of prior knowledge in the absence of an initial consensus (cf. Carlile and Rebentisch 2003).

The value of a digital platform for the end user lies in its orchestrated components (cf. Normann and Ramirez 1993). Each end user actively participates in a knowledge process, through which value is co-created with a slightly different value constellation for each end user. This process requires the active and reciprocal exchange of knowledge between the end user and the platform. The competencies of both the end user and the service provider are used to create value, and knowledge is recreated and transformed in the process (cf. Spender 1996), thus making the interaction between the service provider and the customer a continuous and dynamic event. Boundary objects ease and speed up the interactive communication process. The next section further introduces the role of service artifacts in the value co-creation of digital platforms, followed by examples of real-life service artifacts and their conceptualization based on boundary object theory.

3 Service Artifacts in Digital Platforms

Digital platforms include a personalized web interface for each end user, which Chesbrough (2011) calls a "flexible front end." This interface allows for communication between the platform and the end user. For example, in banking services, end users pay their bills through a web interface that is personalized for each end user; on Amazon.com, each end user sees his/her buying history and recommendations. End users of the UPS package delivery service can track their parcels through a flexible front end. Behind the flexible front end, standardized business processes that are invisible to the end user keep the platform running (Chesbrough (2011)).

In this paper, the concept of service in the context of flexible front ends is approached using the S-D logic perspective. The focus is essentially on value co-creation as defined by S-D logic. When the paradigms of S-D logic are applied to the context of digital platforms, it is clear that flexible front ends provide indirect service in all cases other than direct person-to-person interaction (e.g., chat and other situations with the virtual presence of customer representatives). This type of indirect encounter can also be viewed as technology-mediated self-service (e.g., Bitner et al. 2000; Meuter et al. 2000). In addition, S-D logic's direct service, in which a service is exchanged for a service, is rare in real life and more of an abstract concept. Thus, the flexible front end serves primarily as distribution mechanism, or appliance, for direct service provision (Vargo and Lusch 2007), aiding the service exchange. Therefore, indirect service appliances in digital platforms can be categorized into the same group as all other artifacts, enabling collaboration with the producer of that good and using the knowledge of that producer.

Digital platforms make it possible to facilitate value co-creation in an indirect service setting using diverse and even immersive techniques. The features of indirect service appliances that enable this facilitation mimic the characteristics of direct service provision. The degree of imitation varies according to the use case and the available resources, but the level of facilitation is definitely always greater than in the case of plain "goods." Thus, it can be argued that categorizing indirect service appliances in the flexible front end into the same class as, e.g., tangible goods does not take into account the extensive ways of facilitating the value co-creation between different actors. In other words, the level of facilitation included through the indirect service appliances in the flexible front ends plays so important a role that a transitional concept is needed to assess, analyze, characterize, and categorize different ways of facilitating value co-creation. The transitional concept introduced in this paper is called a service artifact. Figure 3.1 illustrates the position of the service artifact in S-D logic's concept hierarchy.

Figure 3.2 shows where the service artifact is located in the context of the digital platform's flexible front end, between the internal processes and as an integrated part of the personalized front end. The service artifact offers an abstract, dynamic, and S-D logic-compatible way to examine the focal issue related to digital platforms—how value is co-created in a digital platform. The role of the service artifact is to facilitate the value co-creation between the parties in indirect service provision.

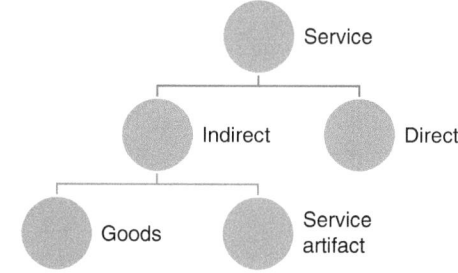

Fig. 3.1 The concept of a service artifact in the S-D logic concept hierarchy

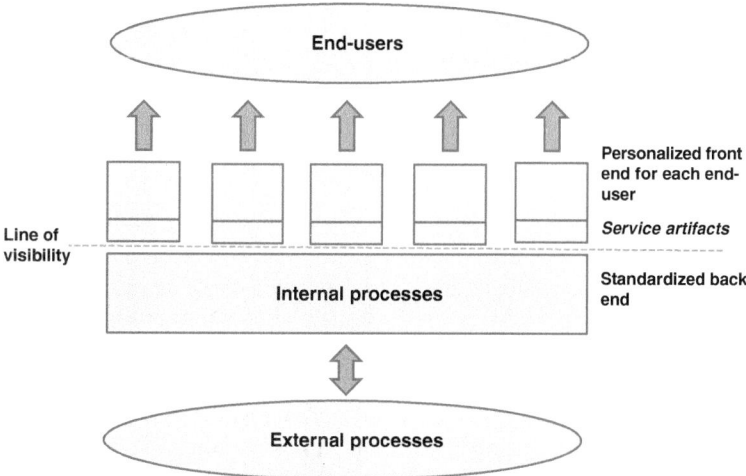

Fig. 3.2 The locus of service artifacts in digital platforms (Chesbrough 2011, modified)

In some cases, only self-service and simple service artifacts are needed. In others, more complex issues may require intelligent service artifacts and greater facilitation. The service artifact can be classified in a continuum, with "goodlike" self-service implementations without specific value co-creation facilitation characteristics on one end and intelligent indirect service instances with active measures of two-way facilitation on the other. Different types of service artifacts can then be placed along the continuum.

The critical notion regarding the service artifact is that it does not belong to any specific customer experience-related or technology infusion-related topics. Most importantly, the term "self-service" as used in this paper refers to the definition in S-D logic (self-service as a case in which the actors themselves operate using appliances) and not to the self-service systems (SST) literature. A service artifact is an independent S-D logic-based transitional concept, enabling an abstract and "pure" S-D logic and service science perspective from which to examine the service and value co-creation in digital platforms. Similarly, the concept of a service artifact is included in all cases of indirect service, regardless of the interactiveness or technological implementation of the user interface. The service artifact covers all indirect service settings in digital platforms—from making simple routine tasks more efficient (e.g., paying invoices and checking the status of an order) to resolving demanding situations in a more customer-oriented way (e.g., video chats enhancing the reach of rich personal encounters). The unit of analysis is the features and the level of value co-creation facilitation in the indirect service setting. Real-life examples provided in the next section further elucidate the concept of a service artifact.

4 Examples of Service Artifacts

The first example of a service artifact is the personalized recommendation lists at Amazon.com. This service artifact provides a means through which the end user becomes aware of new opportunities for consumption. Because the list evolves based on the customer's browsing and buying history, the artifact evolves as well. Amazon.com's recommendation lists work based on data that accumulates in the platform over time. These database-type service artifacts assist the end user to gather relevant information and learn about the opportunities offered by the platform.

The second example is a service artifact integrated by an online banking service (company name intentionally withheld) into its online banking extranet. The central element of this customer service implementation is a question box that allows the customer to freely submit any question. While typing the question, the user is simultaneously shown matching question/answer pairs from a large database of frequently asked questions. In addition, every answer appears with a customer representative face and brief details regarding that person, thus associating the provided information with bank personnel. If the issue is not resolved by one of the pairs, the customer can submit the question, which is then redirected to a real customer service person. The company adopted this service artifact because it enhances the scalability of customer service while retaining the personal nature of the service encounter.

The third illustrative example is the Moodagent playlist generator, which is available as a smartphone or Spotify application. Moodagent uses mathematical models to determine features such as mood, emotion, style, and genre of music tracks. The users can have the application randomly select songs that fit their current feelings or let the application determine the songs based on user's listening history. The selection of music based on moods provides the end user with ease of use for the Spotify platform. The automatic mood-based classifying of music also has professional applications in media productions where music catalogs are needed to serve various occasions.

Some companies have created cartoon characters that communicate with customers and provide a "face" for the company. An example of this type of service artifact is the Japanese department store Daimaru Matsuzakaya's Sakura Panda. The panda is a fictional cartoon character that actively communicates with customers through social media, advertisements, and events. End users identify Sakura Panda with the department stores.

The final example is another character-type service artifact associated with Ruukki, a Finnish steel company. The company has a branded customer service program for roofing customers that primarily operates via the Internet. The service, called "Roof Doctor," provides customers with answers to frequently asked questions, gives product installation and maintenance tips, and shows detailed example videos. All information is presented as though it were written and composed by one specific person. The customer is also encouraged to send questions as though they were actually chatting with a real person rather than just sending e-mails to anonymous customer service representatives. This tool also covers products other than Ruukki's own, and product specialists give interviews to publications and blog about their products. The company benefits from this solution in that this tool makes it easier and more natural for the end customers (who are not steel professionals) to ask specific technical questions.

Over time, characters such as Sakura Panda and Roof Doctor begin to represent the company rather than simply functioning as cartoon characters. Customers are able to form a relationship with Sakura Panda, i.e., to develop feelings about the character, identify the character with the company behind it, and associate memories and life events with the character. Sakura Panda has become the Mickey Mouse or Ronald McDonald of Daimaru Matsuzakaya. At the same time, Sakura Panda allows the department store to inform customers about its new products and services. Similarly, Roof Doctor has become a trusted partner for the end users of steel roof-related services. The service artifact has become a part of the customer's daily lives.

In the future, service artifacts may be characters that are equipped with artificial intelligence. As human-machine interactions become stronger, the end user and service artifact may begin to function as a single entity. The artifact may be able to not only recommend the best options but also to make decisions on behalf of the end user. This ability is possible because during use, data accumulates in the platform from each customer and from all platform participants. By using this data, the service artifact will be able to learn and optimize the use cases for the end users. For example, Apple's Siri is a service artifact with the potential to become artificially intelligent. Similar to IBM's Watson, artificially intelligent service artifacts may utilize natural language processing and similar technologies to access the platform core and complementors on behalf of the end user and to decide the most suitable combinations of products and services.

Service artifacts vary from database-type artifacts to artificial intelligent artifacts. Service artifacts can also support customer service by providing a character and a face for the company's identity. Table 3.1 summarizes the categories of service artifacts.

Database artifacts are simply applications or widgets that utilize data that exists in the platform. Database artifacts provide "quick fixes" for common problems. They can be designed to gather data and keep track of end user behavior in the platform. Character-type artifacts add a reciprocal, communicative aspect to the interaction. End users build their perceptions of the platform through interactions with the character, and the character may also gather feedback and development ideas from the end users. In many cases, character-type artifacts require the involvement of customer service personnel, but much of the communication may be executed through applications. Artificial intelligence artifacts build shared intelligence with the end user, and these two components become inseparable.

A popular way to think about knowledge is to conceptualize it as a hierarchy in which wisdom is built on top of knowledge, and knowledge consists of information that is based on data (e.g., Rowley 2007). Complexity increases as information is transformed into knowledge and wisdom. The usage of data is simple, straightforward, and predictable, but wisdom requires complex processes of interactions between different parts of the system. Similarly, the service artifact categories build semantically on top of each other—character artifacts include elements of database artifacts, and artificial intelligence artifacts include elements from both database and character-type artifacts. The artificial intelligence service artifact is thus the most complex and is able to solve challenging problems on behalf of the end user; in S-D logic's terms, artificial intelligence facilitates value co-creation in a manner that almost reaches the level of direct service.

5   Service Artifacts as Boundary Objects

Service artifacts are boundary objects when they are meaningfully and usefully incorporated into a value co-creation process that engages the platform and the end user (cf. Star and Griesemer 1989). This section links the concept of service artifacts to the theory of boundary objects. The classification presented in the previous section is elaborated to establish which roles the various artifacts fill in the platforms' flexible front ends as they level out the inherent differences in knowledge between the end user and the platform.

Service artifacts act as boundary objects by de-bottlenecking the knowledge processes that occur between the specialized knowledge domains of the platform owner, the complementors, and the end users. They provide a shared method for the co-creation of value within the ongoing interaction. The shared context enables

Table 3.1 Service artifact types and examples

Service artifact type	Functioning logic	Interaction type	Examples	Use cases	Complexity and level of facilitation
Database service artifact	Application that provides personalized recommendations and advice to the end user based on individual end user preferences and big data	One way (end user to the artifact)	Amazon.com lists, Spotify Moodagent	Guiding end user, providing "quick fixes," measuring the efficiency of transactions	
Personal service artifact	Identifiable character that provides the impression of personal service and natural dialogue, building a relationship with the customer	Two way but initiated primarily by the end user	Sakura Panda, Roof Doctor	Reciprocal communication, company identity building, gathering development ideas from end users	
Artificial intelligence service artifact	Intelligent system that adapts to each end user's day-to-day life	Two-way dialogue	Apple's Siri, IBM's Watson	Building shared intelligence, creating and renewing identity and norms together with the end user	

Table 3.2 Service artifacts as boundary objects in the flexible front end of digital platforms

	Database service artifact	Character service artifact	Artificial intelligence service artifact
Interpretative flexibility (i.e., personalized value)	Storing of data that is valuable and end user specific	Building a personal and emotional bond with the end user	Building a strong and symbiotic relationship with the end user
Structuring of process needs and arrangements	Answers to frequently asked questions about the service process	Educating the end user to help him/her understand the service process	Acting and deciding on behalf of the end user in different stages of the service process
Providing dynamics between non-structured and tailored uses of the platform	Tailored recommendations with automated recommendation lists	Communicating with the end user and reminding him/her about his/her options. Engaging the end user with different incentives, e.g., gamification	Analyzing the behavior and expectations of the end user, creating new personalized offerings, and making decisions on behalf of the end user

the co-creation of personalized value for each individual end user, an improved understanding of service processes and an improved understanding of platform offerings. The key to the development of a powerful shared context is the representation of knowledge through the use of service artifacts as boundary objects (cf. Star and Griesemer 1989).

To function as a mechanism of communication at the interface between the digital platform front end and the end user, service artifacts must include three key characteristics of boundary objects, which are identified by Star (2010). First, the service artifact must be able to provide personalized value for each end user individually, i.e., the artifact must include sufficient interpretive flexibility. Second, the artifact must act as a means for the end user to understand the service processes and how to participate in them when using the platform, i.e., it must provide a structure of process needs and arrangements. Third, the service artifact must offer the end user an understanding of all relevant service offerings associated with the platform, i.e., it must establish interactions between the non-structured and tailored uses of the platform. Each of the service artifact types fulfills these three mechanisms in its own way (Table 3.2).

Database service artifacts can provide a faster and more convenient way to fulfill routine end user needs compared to the flexible front end alone. They can also provide tailored recommendations and store data that is valuable for the end user. For example, a healthcare service platform can build a patient history database that could serve an important function for the end user. End users need a reliable source of information for their frequently asked questions, and this type of artifact can provide the necessary answers. Database service artifacts allow end users to conveniently access all gathered and stored information.

Character service artifacts can engender emotions in the end user and form a bond with him/her. The character can become a personal agent for the end user, providing reminders and making daily life easier: for example, planning grocery lists or providing recipe suggestions in a retail platform. Characters can educate the end user about the different steps in the service process and encourage the user to try new functionalities so that the end user better understands and uses the platform offering. The most advanced character service artifacts can also use different incentives and gamified experiences to help the end user recognize his/her latent needs for platform products.

Artificial intelligence artifacts build the strongest and most synergistic relationship with the end user. These artifacts "go deep"; they not only help the end user to understand the different consumption possibilities offered by the platform but can also decide what is best for the end user and act on the user's behalf. Over time, dependencies and obligations are likely to develop between the end user and the artificial intelligence artifact, similar to social relationships with human beings, providing a strong basis for lock-in. Artificial intelligence artifacts are able to analyze end user behavior and expectations to make accurate predictions regarding the end user's future needs.

6 Discussion

Digital platforms intermediating transactions are a relatively new phenomenon in their current format. During the recent decades, platforms have emerged such as Betfair.com that cause radical change in the industry, i.e., driving incumbent service providers out of business and forcing others to rethink their business models. Service artifacts acting as boundary objects in the end user interface have emerged along with the popularity of digital platforms.

The framework and categorization provided in this paper may serve as a guide to help platform owners build strategies for improved end user engagement and retention. It must be noted that service artifacts are not always needed to establish a platform. In iTunes, there are no clear examples of service artifacts and the lack of them does not appear to have hindered the success of iTunes so far. In the iTunes case, the boundary object is the overall user experience of its flexible front end. However, the Spotify.com music service appears to act with different logic, as it is based on service artifacts that recommend music to the end user in addition to providing a baseline user experience. Most of Spotify's service artifacts, such as Moodagent, are created by complementors who are given access to Spotify's interfaces, which results in a more variable and probably also more interesting user experience. Furthermore, service artifacts on their own do not create success. The idea of suggestion lists have emerged in almost every e-commerce site, but Amazon.com still holds the leading position among e-commerce sites, largely due to its superb back-office system, which has earned the trust of its end users, and its scale, which allows it to provide a good price.

A service artifact is a far more complex and multifaceted actor than exemplified by the use of avatars as mascots (Sakura Panda) or of expert systems (the retail bank example in this paper) as stand-alone website features. Service artifacts are active communicators that (1) aim to build a strong relationship with the end user, (2) orchestrate the complementary products and services of the digital platform for the end user, and (3) utilize the back-office processes and accumulated data from the digital platform. There are customer touchpoints in the service process (Meyer and Schwager 2007), and the service artifact is an independent touchpoint that needs its own goals and metrics. The most critical issue is the gap between the artifact facilitated customer service and human customer service—the platform owners must remember that the end user forms the same type of relationship with both actors. Thus, the linkage between these must be carefully designed and clearly communicated to the end user to achieve the desired benefits while avoiding confusion and problems.

Customers form emotional bonds with the brands and products that they use, becoming attached to them (Batra et al. 2012). People have an inherent tendency to connect emotions to nonhuman agents (Epley et al. 2007), as evidenced by the success of virtual pets in general and, for example, Tamagotchi toys in particular. These examples suggest that creating an emotional bond with an end user through a service artifact increases the success of the digital platform. Communication with virtual characters provides, up to a point, experiences of amusement and joy that are similar to the experiences that people have when communicating with living, responsive organisms. Service artifacts may change the role of customer service because they do not require human characteristics to create personal relationships. The role of direct service would diminish and expensive customer service resources could be focused on the encounters that provide the greatest co-creation of value facilitation potential.

Service artifact thinking can be extended beyond virtual communication mechanisms to include physical objects. These may include wearable devices that track and monitor end users, communicating with the platform and providing recommendations on the go. Disney's MyMagic+ wristband or Jawbone devices provide direction for future wearable service artifacts. Possible use cases for these are, for example, those types of services that are not routinely consumed by the user, and the benefit can stem from creating awareness regarding the forthcoming service process. Physical service artifact devices in hospitals could keep the patient updated about queuing times and expected service processes, for example, thus reducing the stress related to the patient's role.

A company of any size can become a platform leader if it is able to create an active ecosystem together with the platform participants (Gawer and Cusumano 2008). The platform owner must build and induce platform processes for efficiency as well as for incremental and radical innovations, if the idea is not to merely act as a mediator in a two-sided market. The complementors' innovation capabilities are needed to improve the core elements of the platform, to improve the quality of the entire ecosystem, and to introduce new offerings. If the complementors do not innovate, the platform may be initially successful in gaining end users' attention, but it may then lose that attention if it fails to evolve to offer sufficient variability for end users. An important strategy is providing technologies that facilitate the complementors' innovation activities (Gawer and Cusumano 2002). Service artifacts could provide the innovation interface.

When a platform is first created, the platform leader faces a chicken-egg problem with the participants (Caillaud and Jullien 2003; Evans 2003): it is difficult to determine which of the participant groups to address first. If service artifacts facilitate innovation activities not only with the end users but also with complementors, then the chicken-egg problem is partially solved. Service artifacts can be fine-tuned to facilitate innovations that improve efficiency and quality and to generate new offerings for end users. Well-designed and executed service artifacts can establish simultaneous communication among all participant groups.

As demonstrated in this paper, service artifacts are a seminal part of digital platforms as boundary objects facilitating value co-creation. Future research into service artifacts should relate to the abovementioned topics such as platform leadership, the definition of the platform core, the orchestration of innovation networked activities and lock-in strategies, as well as chicken-egg problems. The most important questions for the owner of a digital platform are "What are the business benefits of service artifacts?" and "Can service artifacts be used to increase the network externalities of the platform?"

Acknowledgments This paper is the result of research conducted in the User Experience and Usability (FIMECC UXUS) and Future Industrial Services (FIMECC FUTIS) research programs sponsored by the Finnish Metals and Engineering Competence Cluster (FIMECC). The authors wish to thank their colleagues Ms. Hoda Faghankhani and Mr. Otto Mäkelä for their insights.

References

Batra R, Ahuvia A, Bagozzi RP (2012) Brand love. J Market 76(2):1–16

Bitner M-J, Brown SW, Meuter ML (2000) Technology infusion in service encounters. J Acad Market Sci 28(1):138–149

Caillaud B, Jullien B (2003) Chicken and egg: competition among intermediation service providers. RAND J Econ 34(2):309–328

Carlile P, Rebentisch E (2003) Into the black box: the knowledge transformation cycle. Manag Sci 49(9):1180–1195

Ceccagnoli M, Forman C, Huang P, Wu DJ (2012) Cocreation of value in a platform ecosystem: the case of enterprise software. MIS Q 36(1):263–290

Chesbrough H (2011) Open services innovation. Jossey-Brass, San Francisco

Cohen WM, Levinthal DA (1990) Absorptive capacity: a new perspective on learning and innovation. Adm Sci Q 15(1):128–152

Davies M, Pitt L, Shapiro D, Watson R (2005) Betfair.com: Five technology forces revolutionize worldwide wagering. Eur Manag J 23(5):533–541

Eisenmann TR (2008) Managing proprietary and shared platforms. Calif Manage Rev 50(4):31–54

Eisenmann TR, Parker G, Alstyne M Van (2008) Opening platforms : how, when and why? Harvard Business School working papers, 09–030

Epley N, Waytz A, Cacioppo JT (2007) On seeing human: a three-factor theory of anthropomorphism. Psychol Rev 114(4):864–886

Evans D (2003) Some empirical aspects of multi-sided platform industries. Rev Netw Econ 2(3):191–209

Gawer A, Cusumano MA (2002) Platform leadership: how Intel, Microsoft, and Cisco Drive Industry innovation. Harvard Business School, Boston

Gawer A, Cusumano MA (2008) How companies become platform leaders. MIT Sloan Manag Rev 49(2):28–35

Gawer A, Henderson R (2007) Platform owner entry and innovation in complementary markets: evidence from Intel. J Econ Manag Strat 16(1):1–34

Grant RM (1996) Prospering in dynamically-competitive environments: organizational capability as knowledge integration. Organ Sci 7(4):375

Liebowitz SJ, Margolis SE (1994) Network externality: an uncommon tragedy. J Econ Perspect 8(2):133–150

Meuter ML, Ostrom AL, Roundtree RI, Bitner MJ (2000) Self-service technologies: understanding customer satisfaction with technology-based service encounters. J Market 64(3):50–64

Meyer C, Schwager A (2007) Understanding customer experience. Harv Bus Rev 85(2):116

Nonaka I, Takeuchi H (1995) The knowledge-creating company: how Japanese companies create the dynamics of innovation. Oxford University Press, New York

Normann R, Ramirez R (1993) From value chain to value constellation: designing interactive strategy. Harv Bus Rev 71(4):39–51

Prahalad CK, Ramaswamy V (2002) The co-creation connection. Strat Bus 27(2):50–61

Prahalad CK, Ramaswamy V (2004) The future of competition: co-creating unique value with customers. Harvard Business School, Boston

Rowley J (2007) The wisdom hierarchy: representations of the DIKW hierarchy. J Inform Sci 33(2):163–180

Shapiro C, Varian H (1999) Information rules. A strategic guide to the network economy. Harvard Business School, Boston

Spender J-C (1996) Making knowledge the basis of a dynamic theory of the firm. Strat Manag J 17(Winter):45–62

Star SL (2010) This is not a boundary object: reflections on the origin of a concept. Sci Tech Hum Val 35(5):601–617

Star SL, Griesemer JR (1989) Amateurs and professionals in Berkeley' s Museum of vertebrate zoology, 1907–39. Soc Stud Sci 19(3):387–420

Vargo SL, Akaka MA (2009) Service-dominant logic as a foundation for service science: clarifications. Serv Sci 1(1):32–41

Vargo SL, Lusch FR (2004) Evolving to a new dominant logic for marketing. J Market 68(1):1–17

Vargo SL, Lusch RF (2007) Why "service"? J Acad Market Sci 36(1):25–38

Vargo SL, Lusch FR (2008) From goods to service(s): divergences and convergences of logics. Ind Market Manag 37(3):254–259

Vargo SL, Lusch RF, Akaka MA (2010) Advancing service science with service-dominant logic. Clarifications and conceptual development. In: Maglio PP, Kieliszewski CA, Spohrer JC (eds) Handbook of service science. Springer, Boston, pp 134–156

Chapter 4
Four Axiomatic Requirements for Service Systems Research

David Reynolds and Irene CL Ng

Abstract Service science research is a rapidly growing interdisciplinary, cross-functional discipline. As such, it is necessary for formal structures to be created to focus researchers' efforts towards a common end. This chapter starts with an overview of the role of service-dominant logic and systems thinking in service science, lending support to the assertion that service systems should be the 'basic abstraction' of service science research. The chapter then proceeds to argue for four axioms which are necessary to progress knowledge in the domain of service systems.

Keywords Emergence • Holism • Resource integration • Service systems • Service-dominant logic

1 Introduction

We are now over 10 years into the fifth period of service research (2000–present), called the 'Creating Language' period by some researchers (IfM and IBM 2007; Keränen and Ojasalo 2011; Briscoe et al. 2012). In their review of the 2011 Grand Challenge in service conference, held at the University of Cambridge, Briscoe et al. (2012) describe the Creating Language period as the time where new models of service emerge and the concept of service systems develops further, uniting different perspectives within service science. The field is expanding rapidly with increasing numbers of researchers, conferences and networks, while initiatives such as

D. Reynolds (✉)
International Institute for Product and Service Innovation (IIPSI),
The University of Warwick, Coventry CV4 7AL, UK
e-mail: d.j.reynolds@warwick.ac.uk

I. CL Ng
International Institute for Product and Service Innovation (IIPSI),
The University of Warwick, Coventry CV4 7AL, UK

Marketing and Service Systems, Warwick Manufacturing Group (WMG),
The University of Warwick, Coventry CV4 7AL, UK
e-mail: irene.ng@warwick.ac.uk

© Springer Japan 2015
K. Kijima (ed.), *Service Systems Science*, Translational Systems Sciences 2,
DOI 10.1007/978-4-431-54267-4_4

service science management and engineering (SSME), introduced by IBM, aim to strengthen interactions between industry, academia and government (Hefley and Murphy 2008).

Building on the work by Agrawal (2001) and Cronin (2003), Moussa and Touzani (2010) call this period the 'airborne' phase (2004–now). They highlighted the following features of this phase:

- In 2004, service science management and engineering (SSME) emerged as a new interdisciplinary field.
- There was a marked increase in the number of service journals being published.
- A significantly larger proportion of articles in leading marketing and management journals are about service.
- New paradigms and concepts have been developed. Some have had greater impact than others (e.g. the service-dominant logic and the rental/access paradigm).
- There has been increased emphasis on the interdisciplinary, cross-functional and international nature of the field. This includes the integration of computer science, operations research, engineering, management, marketing, social and cognitive sciences and legal sciences.

Service science has gained popularity amongst academics and practitioners as it is seen by some as a way to drive innovation, competition and quality of life through the co-creation of value (Moussa and Touzani 2010; Ostrom et al. 2010).

A variety of fields, traditions and methods are being utilised in this space, from natural and ecological sciences to information technology and cybernetics (Mele et al. 2010). Hence we believe that to move this discipline forward, service science research needs to be structured in such a way as to focus researchers' efforts towards a common end. Few service researchers would disagree with this, and as such, the purpose of this chapter is to clarify some of the key concepts and explore some of the insights gained from what is rapidly becoming a well-developed body of literature.

2   Service-Dominant Logic and Service Systems

Vargo and Akaka (2009) argue that the appropriate foundation for service science research is the service-dominant (S-D) logic, and hence the foundational premises of S-D logic (Vargo and Lusch 2004, 2008; Vargo and Akaka 2009) should form the core of the postulate base (Ng et al. 2012), i.e. adherence to S-D logic is a necessary condition for service science research. The four core foundational premises are summarised in Table 4.1.

As set out in FP1 in Table 4.1, service[1] is the basis of all exchange. In other words, service is always exchanged for service.

---

[1] Singular, indicating a process as opposed to the plural *services*, indicating intangible units of output.

Table 4.1 Core foundational premises of service-dominant logic

Logic premise		Explanation/justification
FP1	Service is the fundamental basis of exchange	The application of operant resources (knowledge and skills), 'service', is the basis for all exchange. Service is exchanged for service
FP6	The customer is always a cocreator of value	Implies value creation is interactional
FP9	All economic and social actors are resource integrators	Implies the context of value creation is networks of networks (resource integrators)
FP10	Value is always uniquely and phenomenologically determined by the beneficiary	Value is idiosyncratic, experiential, contextual and meaning laden

Source: Vargo and Akaka 2009

When service is exchanged, one entity integrates resources, which are 'unique, or otherwise costly-to-copy, inputs' (Conner 1991). A traditional product is therefore a bundle of potential resources proposed to the consumer and service and is defined as the application of the competency afforded by the potential resources that become actual resources to be integrated by the consumer in context.

It can also be argued that it is the integrator that determines whether or not the service was of benefit to them. In other words, *value* can only be created within the mind of the *beneficiary of service* (*consumer*), through the application of their own competencies with those provided by the value proposition in question (Ng 2013). This is known as *value co-creation*.

Value co-creating entities, be they individuals, groups, organisations, firms or governments, are often viewed as interacting with one another within systems, constellations or networks of resources (e.g. (Normann 2001; Normann and Ramírez 1994; Vargo and Lusch 2011; Lyons and Tracy 2013). Each of these systems is an arrangement of resources, connected by a value proposition (Spohrer et al. 2007; Lusch et al. 2008; Maglio et al. 2009; IfM and IBM 2007; Smith and Ng 2012) or more specifically 'service systems'.

Many service researchers therefore turn to systems science for their research, not only because the general systems theory provides the foundation for creating a formal structure of service systems (Maglio et al. 2009; Golinelli et al. 2002)[2] but also because these frameworks exhibit greater robustness arising from their development over some 50 years (Spohrer et al. 2012). Systems science also provides an established lexicon of systems characteristics which can be used to formulate a research agenda for service systems (Ng et al. 2011). These characteristics include: boundaries, interfaces, hierarchy, feedback and adaptation to which most systems writers would add emergence, input, output and transformation (Kast and Rosenzweig 1981; Christopher 2010).

---

[2] For a detailed review of some of the main systems approaches, such as general systems theory (Bertalanffy 1972) and open systems theory (Boulding 1956; Katz and Kahn 1978).

Table 4.2 Recent service system definitions

Service system definitions	Authors	Year
Service systems represent value co-creation configuration of people, technology, value propositions connecting internal and external service systems and shared information	Spohrer, Maglio, Bailey and Gruhl	2007
Service systems can simply be a software application, or a business unit with an organisation, from a project team, a business department, a global division; it can be a firm, institution, government agency, town, city or nation; it can also be a composition of numerous collaboratively connected service systems within and/or across organisations	Qiu, Fang, Shen and Yu	2007
Service systems act as resource integrators, understandable in terms of elements of a work system, within the organisation and through the network enduring resource specialisation, those operand and operant, such as knowledge, skills, know-how, relationship, competences, people, products, money, etc.	Spohrer, Anderson, Pass and Ager	2008
Every service systems is both a provider and client of service that is connected by value propositions in value chains, value networks or value-creating systems	Vargo, Maglio and Akaka	2008
A service system is any number of elements, interconnections, attributes and stakeholders interacting in a co-productive relationship that create value, in which principal interactions take place at the interface between the provider and the customer	Spohrer, Vargo, Maglio and Caswell	2008
Service systems are a complex interplay between firm and customer that form an open system which needs to be designed using the techniques of viable systems and systems dynamics, in which both parties are focused on achieving outcomes	Ng and Maull	2008
	Ng, Maull and Yip	2009
Service systems can be divided into 'front stage' (about provider/customer interactions) and 'back stage' (about operational efficiency), and service performance relies on both of them, putting people (customers and employees), rather than physical goods, in the centre of its organisational structure and operations. The smallest service system is a single person; the largest one is represented by the global economy	Qiu	2009

Source: Barile et al. 2012

The link between a systems science approach and modelling and understanding service has been emphasised by many authors (Barile and Polese 2010; Golinelli et al. 2002; Ng et al. 2012; Briscoe et al. 2012). As seen in Table 4.2, there have also been many attempts to define service systems. However, it was Maglio et al.'s (2009) seminal work which really brought forward the application of systems science to service research. In it, they proposed the service system as the *basic abstraction* of service science. Their definition of a service system, was

"an open system (1) capable of improving the state of another system through sharing or applying its resources (i.e., the other system sees the interaction as having value), and (2) capable of improving its own state by acquiring external resources (i.e., the system itself sees value in its interaction with other systems). In this context, economic exchange depends on voluntary, reciprocal value creation between service systems (each system must willingly interact, and both systems must be improved)." (Maglio et al. 2009)

Despite this attempt at formalisation, they had only just begun the process of abstraction for service science (Maglio et al. 2009). As a result, the stage is set for service researchers to identify and develop exactly how a service system might be investigated.

Some other examples of service system definitions can be found in Table 4.2.

As evident from the table above, there are many different systems approaches which could be applied to service systems (Mele et al. 2010). This also means that not all perspectives are applicable to service systems, as some contradict others. Of course, with so many disciplines to choose from, it is no surprise that there are a large number of frameworks currently being applied.[3] The word 'systems' is liberally used, even when the authors do not subscribe to the basic tenets of systems science. Often, the word 'systems' is used merely to describe the existence of multiple entities in the same space, regardless of what the relationship is between them. To that extent, the rest of this chapter proposes a set of *axioms* as a starting point for how a *service system* should be understood and researched into.

3   Holism

A system can be defined as an entity, which is a coherent whole (Ng et al. 2009), meaning it is not simply the sum of its parts (Godsiff 2010; Mele et al. 2010). This is not to say that the parts are unimportant. Systems thinking instead emphasises the importance of the *relationships* between parts (*entities*) and not the individual parts themselves (Forrester 1958; Ng et al. 2009; Godsiff 2010; Mele et al. 2010). This is known as *holism*.

From the service science literature, there is a feeling that holistic and interconnected approaches are an appropriate starting place for describing service (Godsiff 2010). Some authors go even further and suggest that the whole range of complex human social systems are actually instances of nested, networked *holistic* service systems (Spohrer et al. 2012).

For example, the socio-technical school draws the general conclusion that the social and psychological aspects of work need to be understood in the context of the task and the way in which the technological system as a whole behaves (Emery and Trist 1960). The technology system here is taken to include not only the hardware, machines, etc., but the methods and procedures of work and how that work is organised in a process. Similarly, recent research in socio-materiality challenges the assumption that technology, work and organisations should be conceptualised separately and advances the view that the social and technical (material) worlds are inseparable, or constitutively entangled (Orlikowski and Scott 2009). This means all entities in the system must be considered, regardless of whether human, material or technological.

---

[3] It is not within the scope of this chapter to discuss all the different models of service systems currently being pursued by other service science researchers. It should also be stressed that 'no model of any complex system [like a service system] can be completely right... models are neither right nor wrong. Models are more or less useful...' (Christopher 2010).

Schatzki (2003, 2005) exemplifies this concept of entanglement, proposing that central to all human social interaction are practice-arrangement meshes, where human 'practices' interact with material 'arrangements'. These 'meshes' are similar to service systems in that they 'interlace' in such a way that they build larger and larger 'nets'. Schatzki (2005) uses the example of classrooms linking and overlapping with the department office, college administration offices, dorms, the bookstore and the central administration building at the same 'level' to create the university or a 'practice-arrangement bundle' (Schatzki 2005). This could just as easily be called a service system. The university 'bundle' is tied to other educational institutions, state governments, local city governments, foundations, industries and so on to form the larger 'net' (or service system): American education. As with the socio-technical school, context is key and the emphasis is on the relationship between the human (social) and the material (technical).

It has also been suggested that the strength of largely loosely coupled relationships between entities plays a significant role in both the co-creation of value and formation of service systems (Vargo and Akaka 2012). Hence any model of service systems developed is not purely social, technical or material, but a combination of human, material and technological entities which must be considered according to the connection/distinction and reduction/holistic analysis.

Service systems are therefore *holistic* in nature but could also be *reducible*. The two concepts are not mutually exclusive and ideally, any model of service systems should allow for both the observation of a single entity (reductionism) and a system view of the whole (holism) (Mele et al. 2010; Ng et al. 2011, 2012; Barile et al. 2012). The synthesis of these two approaches is crucial towards understanding both the single element and its relationships with other elements without missing the whole picture and its systemic interpretations. This implies that entities within a system have both a distinctive and a connective role. Hence we argue that a fundamental axiom of a service system is that we cannot choose to consider one role without the other:

First axiom of a service system: Systemic entities must be discussed based on both connective and distinctive roles within the system and how tightly coupled its entities are. Reducibility analysis must therefore report on the implications to the system's holistic nature.

This is consistent with many of the current models being proposed within service systems research. For example, Knowledge Based Service Systems (De Santo et al. 2011)[4] are the convergence of advances in IT tools with the evolution in thinking about system dynamic interactions, adaptive skills, sustainable development, enhanced learning, reconfiguration capacities and service innovation (IfM and IBM 2007) in complex environments (Basole and Rouse 2008).

---

[4] An extension of *smart service systems (SSS)* (Barile and Polese 2010). SSS are greatly concerned with the interconnected nature of the actors in the system. In particular, the relationships between actors may not, at first, be obviously of interest. Proponents of SSS argue that this focus really contributes to the competitiveness of the whole system.

Similarly a work system (Alter 2008, 2012) is defined as a system in which human participants and/or machines perform processes and activities using information, technology and other resources to produce products/services for internal or external customers (Alter 2012).

Despite being clearly based on the 'traditional' goods-dominant logic, even a product-service systems approach (Baines et al. 2007) leads to conceptualising the firms' offering as an integrated view of material (tangibles) and nonmaterial (intangibles) components with the collective aim of fulfilling customer needs (Smith et al. 2012).

4 Emergence

It has already been suggested that service systems can be used to represent a range of complex human systems, including firms, individuals, nations, markets, communities and so on. These are referred to as 'open systems' because they interact with many other systems and exchange resources (e.g. energy, matter and information) (Barile and Polese 2010). An open system suggests a complex and dynamic interaction of the organisation and its environment with undeterminable results (Mills and Moberg 1982), as opposed to a system where no material enters or leaves it (closed). A system is therefore 'open' if it is able to exchange energy, matter and information with its environment (Mele et al. 2010). Hence it can be argued that service systems are often open.

The interactions of the relationships between entities within a service system form a higher-order construct that becomes the driver of value (Lusch et al. 2010), i.e. it is the interaction between entities within the system that drives value and not the entities themselves. Interactions create emergence. An emergent quality is related to the inputs and processes of the system, yet it is unpredictable in the sense that knowing what the individual parts of the system are and how they relate to each other does not necessarily mean one can predict the properties of the whole system (Gummesson 2008; Godsiff 2010). Often, emergence arises from the degree of openness within the system (Bertalanffy 1972).

Ng et al. (2011) provide three insights into this:

a. Organisational life does not often behave in one-way causality, i.e. the elements of the system acting on each other are both changed in some way through their interaction with each other.
b. Emergence is very hard to predict because of the number of elements that interact to produce the property.
c. Not only does the interaction between two elements change them, but often something is produced in the interaction which is 'greater than the sum of its parts' (Ng et al. 2011).

Hence while some service systems that are loosely coupled may not exhibit a high level of emergence, e.g. a bank loan application and approval system, the emergent property must still be reported nonetheless to be consistent with systemic approaches.

Second axiom of a service system: Even while much of the outcome of a service system could be predictable, a service system must exhibit some emergent property.

An example of emergence as applied to value co-creation (and by extension service systems) is how an appreciative system[5] develops a series of 'norms' over time which are responsible for its regulation and are not predictable/programmable (Regev et al. 2011).

5 Perspective and Boundaries

One of the challenges with open systems (and by extension, open service systems) is that due to the easy exchange of resources, it can be difficult to identify what is actually part of the service system and what is just part of the wider environment. In order to deal with this issue, the boundary of the system in focus must be defined.

The boundary is a subjective concept, sometimes called the 'interface' or 'membrane' (Godsiff 2010), which differentiates one service system from another. As service systems are often 'open', the boundary will have points at which two (or more) open systems interact. The definition and interpretation of a given boundary varies according to circumstances. However the definition of a system boundary depends on the 'view point' of the system in focus. This can vary depending on the 'actor in focus' and even applies to entities which are not living, such as firms.

It is often implicitly assumed, when viewing and discussing a service system, that the perspective is that of an outsider looking in, i.e. a positivistic and objective view of the system. However, the same system could be understood and described very differently from every entity within the system (Checkland and Poulter 2006; Alter 2008, 2012). Similarly Regev et al. (2011) note that 'the very "function" of a service is likely to be a subject to debate amongst its stakeholders'.

Each entity's decision process at different points of the service system is different, and every system could have a separate set of boundaries depending on the perspective taken. That said, each entity still invokes abductive, inductive and deductive forms of the entity's descriptive model of the world and the formulation of decision rules (optimal, heuristic, intuitive, irrational) that can be used for determining a decision (Ng et al. 2012).

Hence a service system may exhibit outputs that could be both deterministic (predictable) and emergent due to the nature of the interactions between decisions made and the level of autonomy between the entities. The more autonomous the entity, the less likely the outputs will be predictable. So, for there to be a consistent analysis of a service system, the perspective and boundaries of the system must be made clear from the outset.

---

[5]A specialisation of general systems thinking proposed by Vickers (1968) as a way to model how humans and organisations understand and act on their environment.

Third axiom of a service system: The boundaries and perspective of a service system must be specified and held consistent across all discussions.

As a case in point, the definition of the system boundaries is essential when adopting the viable systems approach (VSA) (Golinelli et al. 2002), due to the recursive nature of the model.[6] By introducing the need to report on perspective and boundaries, the third axiom would naturally lead to the role and function of the system, the scope of what it is for and whom it serves from whichever perspective. This echoes Schatzki's (2005) 'site ontology' in that an entity (or event) is tied to context and equally context is tied to the entity (or event). Neither one can exist without the other.

6 Resource Integration and Competencies

All social and economic actors are resource integrators, which are capable of contributing to value co-creation (Barile et al. 2010; Barile and Polese 2010; Vargo and Akaka 2012). Similarly, it has been argued that entities involved in service provision act as integrators of various resources (such as knowledge, skills, know-how, competencies, material resources, money and so on) (Maglio et al. 2009).

Chandler and Vargo (2011) emphasise how social contexts influence, and are influenced by, value co-creation processes within and amongst systems of service exchange. Models of resource integration must define the dynamic and context-specific configurations of form, time, place and possession of resources that achieve the 'density' that is necessary for optimal value creation (Lusch et al. 2010). Density, as defined by Normann (2001), is a measure of the amount of information, knowledge and other resources (e.g. institutions) that an actor has at any given time and/or place to solve problems. Therefore a service system co-creates value for a specific actor through the integration of resources and the availability of potential resources specifies the density of the context.

Since service is an application of competency through which an entity integrates resources to co-create value, competencies of entities within the system could be described through their agencies (capacity of an individual to act independently and to make their own free choices), if human, or their affordances (the quality of something that allows an actor to perform an action upon it), if material (Ng 2013). The decision to act requires judgement, which is based on the context of the system (Ng 2013). In other words, resource integration and competencies arise (and value is created) when agencies take effect in practices and affordances are enacted upon to achieve the systemic outcome, within a specific context.

---

[6]This is an extension of the viable systems model (VSM) (Beer 1984) which has been applied to political systems of nations, pharmaceutical companies, electricity companies and SMEs (Vidgen 1998; Achterbergh and Vriens 2002; Hoverstadt and Bowling 2002; Schwaninger et al. 2004; Haslett and Sarah 2006).

Fourth axiom of a service system: A service system must report the competency (i.e. ability to render the service) of entities within the system.

For example, Alter (2008, 2012) emphasises the need for service to benefit someone (or something), not just the provider of the service. This includes the provision of resources that others will use. Smart service systems (SSS) (Barile and Polese 2010) are 'smart' because they change the way resources are utilised to reflect a change in their environment (Barile and Polese 2010).[7] Similarly, Knowledge Based Service Systems do not only describe the relationships between entities in a system but also identify and classify the resources employed in the process of services exchange between entities (De Santo et al. 2011).

7   Conclusion

We argue that these four axioms, grounded in systems science and S-D logic, are necessary to progress knowledge in the domain of service systems. They will serve to ensure consistency in elucidating implicit assumptions of service systems research and development. Further research is needed to develop models of service systems based on these axioms. We anticipate that the development of predictive models will be particularly challenging. Identifying all of the potential outcomes where there are many interacting elements (all the potential states of the system), and taking into consideration the non-linear relationships and multiple potential feedback loops, means that the results may well be impossible to predict. Hence any model of service systems which claims to be predictive is likely to be very complex (Ng et al. 2011).

Just as the 'winners' of the industrial revolution were the firms who were able to make 'things' for their customers faster and, more efficiently, the 'winners' of the digital revolution will be the firms who are able to serve their customer needs better, we believe, as do many others, that the Art and Science of Service Systems will provide the necessary tools to do this.

References

Achterbergh J, Vriens D (2002) Managing viable knowledge. Syst Res Behav Sci 19(3):223–241
Agrawal M (2001) Building a new academic field in India: the case of services marketing. J Serv Res 1(1):104–120
Alter S (2008) Service system fundamentals: work system, value chain, and life cycle. IBM Syst J 47(1):71–85

---

[7]This is made possible through the application of information and communications technology (ICT) with significant emphasis on automation, or minimal human involvement. Ideally, SSS are capable of self-reconfiguration in order to deliver the best possible performance to satisfy all actors in the system (Barile and Polese 2010).

Alter S (2012) Metamodel for service analysis and design based on an operational view of service and service systems. Serv Sci 4(3):218–235

Baines TS et al (2007) State-of-the-art in product-service systems. Proc IME B J Eng Manufact 221(10):1543–1552

Barile S, Polese F (2010) Smart service systems and viable service systems: applying systems theory to service science. Serv Sci 2(1–2):21–40

Barile S, Spohrer J, Polese F, (2010) Editorial column-system thinking for service research advances. Serv Sci 2(1–2):i–iii

Barile S, Saviano M, Polese F, Di Nauta P (2012) Reflections on service systems boundaries: a viable systems perspective. Eur Manag J 30(5):451–465

Basole RC, Rouse WB (2008) Complexity of service value networks: conceptualization and empirical investigation. IBM Syst J 47(1):53–70

Beer S (1984) The viable system model: its provenance, development, methodology and pathology. J Oper Res Soc 35:7–26

Bertalanffy LV (1972) The history and status of general systems theory. Acad Manage J 15(4): 407–426

Boulding K (1956) General systems theory—the skeleton of science. Manag Sci 2(3):197–208

Briscoe G, Keränen K, Parry G (2012) Understanding complex service systems through different lenses: an overview. Eur Manag J 30(5):418–426

Chandler JD, Vargo SL (2011) Contextualization and value-in-context: how context frames exchange. Market Theor 11(1):35–49

Checkland P, Poulter J (2006). Learning for action: a short definitive account of soft systems methodology and its use for practitioner, teachers, and students (Vol. 26). Wiley, Chichester, UK

Christopher, WF (2010) From system science-a new way to structure and manage the company for sustainable success. Ser Sci 2(1–2):62–75

Conner K (1991) A historical comparison of resource-based theory and five schools of thought within industrial organization economics: do we have a new theory of the firm? J Manag 17(1):121–154

Cronin JJ Jr (2003) Looking back to see forward in services marketing: some ideas to consider. Manag Serv Qual 13(5):332–337

De Santo M, Pietrosanto A, Napoletano P, Carrubbo L (2011). Knowledge based service systems. In Proceedings of the 2011 Naples Forum on Service: Service Science, SD logic and network theory Napoli: Giannini

Emery FE, Trist EL (1960) Socio-technical systems. In: Churchman C (ed) Management sciences, models and technique. Pergamon, London

Forrester JW (1958) Industrial dynamics—a major breakthrough for decision makers. Harv Bus Rev 36(4):37–66

Godsiff P (2010) Service systems and requisite variety. Serv Sci 2(1–2):92–101

Golinelli GM, Pastore A, Gatti M, Massaroni E, Vagnani G (2002) The firm as a viable system: managing inter-organisational relationships. Sinergie 58:65–98

Gummesson E (2008) Quality, service-dominant logic and many-to-many marketing. TQM J 20(2):143–153

Haslett T, Sarah R (2006) Using the viable systems model to structure a system dynamics mapping and modeling project for the Australian taxation office. Syst Pract Action Res 19(3):273–290

Hefley B, Murphy W (2008) Service science, management and engineering. Springer, New York

Hoverstadt P, Bowling D (2002) Royal Academy of Engineering. Systems engineering workshop modelling organisations using the viable system model

IfM, IBM (2007). Succeeding through service innovation: a discussion paper, Cambridge, United Kingdom: University of Cambridge Institute for Manufacturing

Kast FE, Rosenzweig JE (1981) General systems theory: applications for organization and management. J Nurs Adm 11(7):32–41

Katz D, Kahn RL (1978) The social psychology of organizations, IIth edn. Wiley, New York

Keränen K, Ojasalo K (2011). Value co-creation in b-to-b services. In: Campus encounters – bridging learners conference "developing competences for next generation service sectors" Porvoo, Finland

Lusch RF, Vargo SL, Wessels G (2008) Toward a conceptual foundation for service science: contributions from service-dominant logic. IBM Syst J 47(1):5–14

Lusch RF, Vargo SL, Tanniru M (2010) Service, value networks and learning. J Acad Market Sci 38(1):19–31

Lyons K, Tracy S (2013) Characterizing organizations as service systems. Hum Factors Ergon Manuf Serv Indust 23(1):19–27

Maglio PP, Vargo SL, Nathan Caswell N, Spohrer J (2009) The service system is the basic abstraction of service science. Inform Syst E Bus Manag 7(4):395–406

Mele C, Pels J, Polese F (2010) A brief review of systems theories and their managerial applications. Serv Sci 2(1–2):126–135

Mills PK, Moberg DJ (1982) Perspectives on the technology of service operations. Acad Manage Rev 7(3):467–478

Moussa S, Touzani M (2010) A literature review of service research since 1993. J Serv Sci 2(2):173–212

Ng I (2013) Value & worth: creating new markets in the digital economy. Innovorsa, Cambridge

Ng I, Maull R, Yip N (2009) Outcome-based contracts as a driver for systems thinking and service-dominant logic in service science: evidence from the defence industry. Eur Manag J 27(6): 377–387

Ng I, Maull R, Smith L (2011) Embedding the new discipline of service science. In: Demirkan H, Spohrer JH, Krishna V (eds) The science of service systems. Springer, New York

Ng I, Badinelli R, Polese FD, Nauta P, Löbler H, Halliday S (2012) S-D logic research directions and opportunities: the perspective of systems, complexity and engineering. Market Theor 12(2):213–217

Normann R (2001) Reframing business: when the map changes the landscape. Wiley, West Sussex

Normann R, Ramírez R (1994) Designing interactive strategy: from value chain to value constellation. Wiley, Chichester

Orlikowski W, Scott S (2009) 10 sociomateriality: challenging the separation of technology, work and organization. Acad Manag Ann 2(1):433–474

Ostrom AL, Bitner MJ, Brown SW, Burkhard KA, Goul M, Smith-Daniels V, Demirkan H, Rabinovich E (2010) Moving forward and making a difference: research priorities for the science of service. J Serv Res 13(1):4–36

Regev G, Hayard O, Wegmann A (2011) Service systems and value modeling from an appreciative system perspective. In: Second international conference on exploring services sciences, IESS. Geneva

Schatzki T (2003) A new Societist social ontology. Philos Soc Sci 33:174–202

Schatzki TR (2005) Peripheral vision: the sites of organizations. Organ Stud 26(3):465–484

Schwaninger M, Ríos JP, Ambroz K (2004) System dynamics and cybernetics: a necessary synergy. In Proceedings: International system dynamics conference. Oxford, UK pp 1–19

Smith L, Ng I (2012) Service systems for value co-creation, In Haynes, K, Grugulis, I (eds) (2013). Managing services: challenges and innovation. Oxford University Press, Oxford, UK

Smith L, Maull R, Ng I (2012) Servitization and operations management: a service-dominant logic approach. Int J Oper Prod Manag 34(2):242–269

Spohrer J et al (2007) Steps toward a science of service systems. Computer 40(1):71–77

Spohrer J, Piciocchi P, Bassano C (2012) Three frameworks for service research: exploring multilevel governance in nested, networked systems. Serv Sci 4(2):147–160

Vargo SL, Akaka MA (2009) Service-dominant logic as a foundation for service science: clarifications. Serv Sci 1(1):32–41

Vargo SL, Akaka MA (2012). Value cocreation and service systems (re) formation: a service ecosystems view. Serv Sci 4(3):207–217

Vargo S, Lusch R (2004) Evolving to a new dominant logic for marketing. J Market 68(1):1–17

Vargo SL, Lusch RF (2008) Service-dominant logic: continuing the evolution. J Acad Market Sci 36(1):1–10

Vargo SL, Lusch RF (2011) It's all B2B…and beyond: toward a systems perspective of the market. Ind Market Manag 40(2):181–187

Vickers G (1968) Value systems and social process. Tavistock, London

Vidgen R (1998) Cybernetics and business processes: using the viable system model to develop an enterprise process architecture. Knowl Process Manag 5(2):118–131

Chapter 5
Social Innovations—Manifested in New Services and in New System Level Interactions

Marja Toivonen

Abstract This chapter builds bridges between the research areas of service, social, and system innovations. It highlights the need for an integrated perspective in order to answer the big challenges of today's society and to exploit the opportunities provided by smart technologies. Several approaches applied in the research into service innovations are also relevant in the context of social and system innovations, but broadening the scope from the provider-customer dyad to a multi-agent framework is necessary. Collaborative practices play a crucial role and particular attention has to be paid to dissemination of innovations, in addition to the efforts of creating them.

Keywords Empowerment • Open innovation • Social innovation • Systemic issues • User-driven practices

1 Introduction

Since the mid-1990s, research into service innovation has rapidly accumulated. Three main approaches can be identified in this research. First, *quantitative innovation surveys* have been used to identify the generality of innovation activities in various service sectors, and *new indicators* suitable to recognizing service innovations have been developed (e.g., Kuusisto et al. 2011; Rubalcaba et al. 2010). Human resources as an important form of innovation expenditures have been highlighted in this context, and the linkages between service innovations and organizational innovations have been emphasized (van der Aa and Elfring 2002).

Second, *innovation in services has been modeled from both the process and outcome perspectives*. The former efforts have typically adopted the traditional R&D process as an ideal (e.g., Alam and Perry 2002), but also more experiential process models have been suggested (Engvall et al. 2001; Toivonen 2010). The outcome perspective includes the modeling of a service product (offering) in a way that enables the identification of its novel elements resulting from innovation. The most famous model of this type is the model of Gallouj and Weinstein (1997) which

M. Toivonen (✉)
VTT Technical Research Center of Finland, Espoo, Finland
e-mail: Marja.Toivonen@vtt.fi

© Springer Japan 2015
K. Kijima (ed.), *Service Systems Science*, Translational Systems Sciences 2,
DOI 10.1007/978-4-431-54267-4_5

describes a service as a set of final characteristics (user benefits), technical characteristics (production systems), and competence characteristics and defines service innovation as any change in these characteristics.

Third, *the ways to foster innovation activities in services at the organizational level* have been searched for. The "balanced empowerment" system presented by Sundbo (1996) has been the basis for the current interest in employee-driven innovation, which in the newest studies has also been combined with the perspective of user-driven innovation (Hasu et al. 2011). "Balanced empowerment" is a general innovation system involving most employees in an organization. The task of the management is to inspire innovations on the basis of the organizational strategy but also define the framework within which the innovations should be kept (Sørensen et al. 2013).

While research in the above-described areas has been useful, the need for a more holistic stance has become apparent during recent years. The current social, economic, and environmental challenges are too big to be solved via individual product and service innovations created in individual organizations. A crucial question is how to combine various innovations effectively and disseminate them rapidly on the basis of continuous interaction of different organizations. In other words, examining and developing *innovations at the systemic level* have come to the fore. The approaches applied in service innovation research form a good starting point for the structuration of research at this broader level, too. We can apply indicator approaches to map best practices in the development of system innovations in various countries and regions. We can also build up models that describe the nature of system innovations and the processes in which they emerge. Finally, we can construct models that describe the fostering and management of innovation activities in the multi-agent interaction involving various organizations.

System innovations are interlinked with social innovations. The concept "social" includes two different aspects that are both essential when innovations are pursued at the system level. First, *the prominent challenges are societal*, concerning environmental and social sustainability in the first place: energy consumption, climate change, aging, unemployment, and social exclusion. These challenges require new solutions in the areas of community infrastructures, housing, workplace design, healthcare, education, etc. Second, "social" refers to *the participatory and networked processes* without which it is not possible to create innovations in a multi-agent environment. While the challenges that we face today are big, there are also new opportunities for solving them via "smart growth," based on the effective interplay of various knowledge sources via ICT. A prerequisite for the realization of this opportunity is, however, that various stakeholders engage actively in the creation, implementation, and diffusion of innovations.

Social innovations pose *new requirements to policy makers*. At the macro level, there is a need to enhance society's innovation capacity. At the meso level there is a need to revitalize innovation institutions and foster the innovation activities of public, private, and third sector organizations. At the micro level there is a need to ensure that innovations engage with and are driven by the aspirations of communities and individual citizens (Rubalcaba et al. 2011). All these activities necessitate a strong and coherent nexus between the "knowledge triangle" of education, research, and innovation and the development and monitoring of new policies. A new emphasis is the empowerment of citizens: their role is not a passive recipient of innovations shaped by others, but an active cocreator in the innovation process.

This chapter aims to summarize the state of the art in the research that combines the perspectives of service, social, and system innovations. It starts by opening up the concept and central topics of social innovation and thereafter analyzes the relationships between social and system innovations. The perspective of service innovation is involved throughout, because most social innovations manifest themselves as new services. We also supplement our analysis with a review on two neighboring research fields: user-driven innovation and open innovation. Results of these fields can be utilized in the further development of studies on social and system innovations.

2 The Characteristics of Social Innovations

As the research into social innovations is only beginning, a detailed and generally accepted definition for the concept is difficult to find. In social innovations, solutions are typically sought for a wide range of issues, representing different realms of society: labor market, education, health, housing, etc. (Moulaert et al. 2005). Their common characteristic is that they concern *complex economic and social problems*. The outcomes of innovation usually arise in the form of a service innovation which benefits the members of a community or the whole community (Harrison et al. 2010).

On the other hand, researchers have highlighted the nature of the innovation process as an important characteristic of social innovations besides their content. Here social innovations deviate from service innovations: the interactions taking place *comprise much more than a traditional service relationship*. The sources and goals of innovation are more diverse and the activities and actors more multiple, reflecting the multifaceted nature of social innovations. The participation of actors often includes some voluntary elements (combined with commitment). Social innovations may (1) emerge at the grassroots level among individual citizens who respond to pressing social problems, (2) be produced by private, public, and third sector organizations separately or in cooperation, or (3) result in fundamental changes at the societal and policy level (Rubalcaba et al. 2011). Research in these three areas has focused on the following topics, respectively: the empowerment of citizens and stakeholders, the public-private partnerships and the so-called social economy, and the governance and management of social innovations.

The discussion on *empowerment* highlights that social innovations combine two aspects of social life: the economic aspect and the social aspect. Thus, the aim is not only the production of services and the creation of wealth but also the promotion of values and initiatives involving individual and collective empowerment and the development of democracy and responsible citizenship (Harrison et al. 2010). The process of creation and implementation of social innovations relies on *participatory dynamics*, which requires active input from the various stakeholders and results in fostering and utilizing the citizens' social capital in life and work (Nahapiet and Ghostal 1998). As a research field, studies on participatory practices and empowerment are linked to studies on user- and employee-driven innovation, which is an area of growing interest.

The active role of citizens and their communities is a new emphasis in innovation research. The introduction of social innovations has also changed our notions on more traditional innovation activities—those taking place within and between firms and public organizations. This point of view has focused on *new types of organizations and on the integration of initiatives in existing organizations* (Moulaert et al. 2005). Research has been active concerning the third sector (the so-called social economy) in particular. Here, the noneconomic aspects of economic interventions— e.g., the social integration of disadvantaged people—have been emphasized as an important aspect of the concept "social." Innovations in this context are sometimes called "pure social innovations" because they address needs that are not satisfied through the market mechanism due to the lack of profit potential. The social economy consists of nonprofit organizations (NPOs), cooperatives and associations, social entrepreneurs, and partnerships between the public and third sectors. Social innovations may be produced either autonomously by the third sector or with the public support; a partnership with the public sector is also possible. In the partnerships, the role of the actors of the third sector may vary from that of a subcontractor to common design and implementation of social policies with the public stakeholders (Harrison et al. 2010).

Also private firms are entering the field of social innovation; corporate social responsibility and concern on sustainable development are more and more often a part of their strategies (Lapointe and Gendron 2004). The way in which *the striving for social innovations changes innovation processes* concerns all types of organizations—both public and private. Unlike innovations in the market sector, which traditionally have been kept outside competition as long as possible, *social innovations call for imitation and diffusion.* In them, open innovation is not an alternative strategy but the primary strategy, i.e., forming alliances and networks is essential. The governance and management of these networks have to support both the creation and dissemination of innovations. Dissemination is a challenging task due to two characteristics of social innovations: *local nature and the lack of codification.* The contribution of social innovations is typically manifested as the density of local networks and as local vitality that may result in new jobs and market activities. Scaling up innovations from this limited context requires the strengthening of their systemic features. It also requires new types of R&D practices that can facilitate the codification of social innovations and the procedures applied (Harrison et al. 2010). An interesting approach developed for scaling up social innovations is *societal embedding* (Kivisaari et al. 2013), which focuses on innovation networks with flexible compositions.

3   The Relationship of Social and Service Innovations to System Innovations

The central role of networks in social innovations depicts their interlinkage with system innovations. A system innovation refers to a new operational model which is based on *the simultaneous development of organizations, technologies, services, and multiple network relationships.* An important characteristic of system innovations is that the novelty is not restricted to the ways of operating, but *also the knowledge sources and the ways of interacting with other actors are new* (cf. Gallouj et al. 2013). This aspect points out the various forms of knowledge included in innovation: "knowing who" is essential besides "knowing what" and "knowing how" (Lundvall and Johnson 1994).

Several researchers have highlighted the complexity of system innovations. An important source of complexity is the fact that it is not possible to identify systemic problems directly, but they manifest themselves in various issues of everyday life—often as a service failure. According to Windrum (2008), system innovations have actually much in common with "conceptual innovation": they question the existing knowledge and assumptions that maintain current services, processes, and organizations. In order to create innovations in this kind of a context, a dialog is needed between the conceptual and practical levels. The approach of expansive learning has suggested a way in which this dialog can be carried out. Here, visible problems form the starting point from which the analysis must proceed to the identification of systemic contradictions. New conceptual solutions should then be sought for diminishing these contradictions. Finally, the new solutions should again be concretized so that they can be tested at the practical level in order to see whether they answer the original problems. This stage often includes a renewal of existing services or the introduction of new services (Fig. 5.1).

System innovations can be either business innovations or public innovations or they may concern both realms of society. The concept of *ecosystem* is increasingly used in the analysis of the development of business sectors (Iansiti and Levien 2004). Examples of system innovations including both the private and public sectors are intelligent traffic and intelligent energy systems (e.g., smart grids). They can be combined to be part of an even more comprehensive type of renewals: so-called city

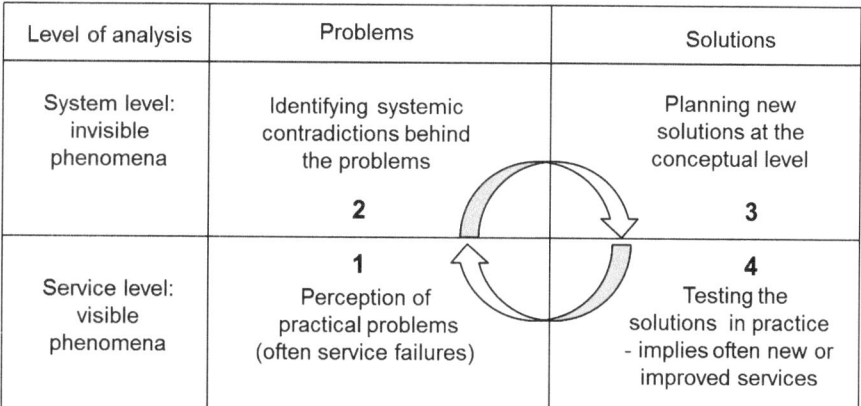

Level of analysis	Problems	Solutions
System level: invisible phenomena	Identifying systemic contradictions behind the problems **2**	Planning new solutions at the conceptual level **3**
Service level: visible phenomena	**1** Perception of practical problems (often service failures)	**4** Testing the solutions in practice - implies often new or improved services

Fig. 5.1 Services as manifestations of problems and solutions in the context of system innovations (modified from Hill et al. 2007)

innovations which combine smart infrastructure with new information systems and novelties in social systems.

In the public sector, the interlinkage of system and social innovations is apparent. Here the complexity derives from the fundamental changes required. The changes concern delivery systems and services, organizational structures and processes, attitudes and values, as well as strategies and policies. The change of values has been emphasized in particular. Harrison et al. (2010) identify three dimensions in social and system innovations: (1) social dimension, strengthening the social links, (2) economic dimension, producing wealth, and (3) political dimension, demand-based actions and the democratization of socioeconomic life. These dimensions can be crystallized into the requirements of *valid empowerment, effective services, and legitimate governance*. Preconditions for their realization are the growth of nongovernmental organizations, new values and beliefs in civil society (participation, autonomy, and empowerment), the presence of strong networks and social movements, and the existence of institutions that can diffuse innovations.

An important point to be taken into account is the dual structure that is inherent in all social systems: they include an informal, loosely coupled interaction structure among people and a formal management structure which expresses the official goals, norms, and values of the system (Giddens 1987). Social innovations require interaction between these two systems and are challenging from the viewpoint of governance and management as they include ambiguous, even contradictory features. They encompass initiatives to promote social cohesion but also movements protesting against the established order. They need managerialist approaches in order to result in efficient and effective services but also approaches that emphasize grassroots initiatives (Harrison et al. 2010).

Currently, there is an ongoing change in the intervention strategies of public management which reconstructs its responses to economic and social crises, weakened social links, and the challenges of welfare state (Harrison et al. 2010). The need to foster learning and innovation in a changing environment has led to the development of new organizing principles in public administration that now evolve in parallel with bureaucracy and market imitating views of "customership." Several researchers refer to a shift from "government" towards "governance": the rise of networks and partnerships, innovations in democratic practice, and the development of coproduction as a service model. Hierarchically organized, unitary systems that govern by means of law, rule, and order are replaced to some extent with more horizontally organized and relatively fragmented systems that govern through the regulation of self-regulating networks (Newman and Clarke 2009; Sørensen 2002).

All this means that social and system innovations do not emerge without policy measures and governance structures that support their creation. In addition, there is urgent demand for the development of practices of *innovation management* for social and system innovations. Innovation management is equally important in this context as in the context of market-based innovations, and its practices can be either *top-down or bottom-up*. There are three main ways in which social and system innovations can be managed on the basis of top-down principle: regulation-based management, management via the allocation of resources and delegation of decision power, and political management. Typically, all of these factors function today as both driving forces and hindering factors of innovation, depending on the specific situation. In the bottom-up management, innovation can be promoted via user-driven practices and via the fostering of open innovation. Openness is essential also in intraorganizational practices in order to efficiently utilize the expertise of employees—collaboration across sectors and professions is a key question here.

4 Neighboring Research Fields: User-Driven and Open Innovation

The embryonic stage of the research into social and system innovations makes it important to link this research with neighboring scientific fields, whose results can be utilized to supplement it and to promote its further development. The fields of user-driven innovation and open innovation are particularly interesting in this respect. The former has old roots but has become more well known during recent years. The latter is a new approach that in a short time frame has provided important insights about the alternatives in innovation activities and in the management of innovation.

4.1 User-Driven Innovation

User-driven views are closely linked to social innovation since they examine social agents as coactors in innovation. In addition, studies in this field have recently developed to directions that broaden the analysis from the provider-customer dyad to the societal context of using products and services and to multiple roles of users (consumers, citizens, etc.).

Understanding the users as a source of innovation is not new. As innovation in general, also the role of users was first theorized in the context of material products. Since the early studies, this role has been understood in two main ways: *taking user needs as the starting point* and *relating to users as innovators.* The former can be traced back to the emergence of interest in "user feedback" in the late 1970s (Nelson and Winter 1977). The latter is based on the studies of von Hippel (e.g., 1986, 2005), whose basic argument has been that users provide more than an idea for a new product. They may supply an innovating firm with the identification of a problem, product-related specifications, or even a product design. Lead users are particularly important as they face needs months or years before the greater part of market encounters them.

In services, the development of corresponding views started within the school of service marketing, which applies the new service development (NSD) framework for the analysis of innovation (Carlborg et al. 2013). Here, the focus is on the relationship between the provider and the customer, and the concept of user is applied rarely.

Service marketing scholars have played a central role in developing managerially oriented research on the question of how a producer acquires and structures information of user needs (e.g., Edvardsson et al. 2006). Customer interface as an arena for the acquisition of versatile understanding has been highlighted besides surveys, which have first and foremost mapped the satisfaction of customers. In the newest studies, two additional perspectives have come to the fore: the role of user experience (Payne et al. 2008) and the importance of elaborating information on user needs into shared understanding within the provider organization (Nordlund 2009).

Research into user-driven innovation is relevant from the viewpoint of social and system innovations due to *its linkages to the issues of bottom-up innovation practices and bottom-up innovation management*. More specifically, studies on user-driven innovation have theorized and modeled *collaboration with users during the innovation process*. These theories and models are applicable in the context of social and system innovations as well, with some modifications and supplementations. Application possibilities can also be found in the views that emphasize *the contextual nature of using products and services*. These views highlight the network perspective and the importance of social relationships, which are core elements in social and system innovations.

Collaboration with users has been analyzed before, during, and after the innovation process. The studies of von Hippel represent the "before" approach: here a user starts the development. The alternative in which the actual innovation process is carried out together with users has gained the broadest attention, and three different applications can be identified in it. First, some researchers have focused on user input in the front-end of innovation which allows greater creativity than the actual development stage (Koen et al. 2001). Second, the traditional stage-gate innovation model (consisting of idea generation, screening, commercial evaluation, detailed development, testing, and commercialization) has been modernized into a model where the input from users can be taken in at every stage (Alam and Perry 2002). Third, some researchers have highlighted the demanding nature of the transfer from development to implementation and considered that the involvement of users in piloting is most crucial (Hasu 2001).

A view that recently has aroused particular interest is "after innovation" (Tuomi 2002; Sundbo 2008). It emphasizes that an innovation does not stay the same throughout its diffusion, but is modified in use. Novelties are interpreted and appropriated by the users, and one novelty has different meanings for different user groups. Furthermore, *social practices change together with the incorporation of new products and services*. These perceptions have led to questioning the ideal of strong preplanning. Instead of it, they favor rapid implementation of ideas in a preliminary or small-scale form. This enables user involvement and the creation of real-time shared experience of the object to be developed (Engvall et al. 2001; Toivonen 2010). "After innovation" and rapid implementation are relevant approaches from the viewpoint of social and system innovations, which typically are practice-oriented and link together development and practice (cf. Harrison et al. 2010).

The contextual and dynamic nature of using products and services is a phenomenon that many researchers with different focuses have recently highlighted.

For instance, researchers examining the experiential side of products and services have pointed out the significance of *social networks as the framework for experiences* (e.g., Payne et al. 2008). The perspective of service-dominant logic (SDL) has raised to the discussion the role of users in value creation. Vargo and Lusch, the developers of this perspective, argue that the value of service is always cocreated by the provider and the user; the provider cannot deliver value on behalf of the user. This is due to the fact that the multiple relationships in the user's economic and social context contribute to the value creation—*the user integrates contextual resources* with the specific input received from the provider (Vargo and Lusch 2004).

Recent research has also emphasized the multiplicity of user groups: ordinary users, critical users, and nonusers, in addition to lead users. Besides individual users, user communities play a growing role as sources of innovation—both existing communities and new communities that grow around novelties (Kaasinen et al. 2010). The research on users is integrating with consumer research, and the *cultural consumption theory* has apparent linkages to the framework of social innovations. It draws attention to social, cultural, moral, and political values that influence individual consumers and consumer groups (Gabriel and Lang 2008). Interesting is also the research into *the interplay between the roles of customer and citizen*. Scholars in this area have highlighted that the rights and responsibilities of citizens are very different from those of customers: citizens are responsible members of a collective, and they are not always sovereign actors but restrained by existing structures, e.g., power structures (Rosenthal and Peccei 2007).

4.2 Open Innovation

Open innovation refers to the use of purposive inflows and outflows of knowledge to accelerate internal innovation and to expand the markets for external use of innovation, respectively (Chesbrough 2011). It is increasingly evident that organizations do not possess all the valuable knowledge in-house, which highlights the utilization external sources. The literature on open innovation focuses on the role of interactive structures and processes, covering a range of more or less formalized cooperation models.

Wang et al. (2012) argue that open innovation represents a quantum leap with respect to the previous literature on collaborative innovation strategies:

- It emphasizes that innovating organizations have to make full use of both internal and external innovations. The idea that external sources of innovations are as important as internal ones was not present in previous literature.
- It offers a unified framework in which an organization's innovation strategy, the choice between external technology sourcing modes, the creation of absorptive capacity, and business model thinking are tightly linked to each other.
- The buzz on open innovation has triggered many firms and organizations to redirect their innovation strategy in new ways.

Table 5.1 Comparing open innovation with social and system innovations (Kuusisto and Vänskä 2011)

	Open innovation	Social and system innovations
Focus and outcome of innovation	Product and technology dominant	Usually intangible in nature and often manifest themselves in service innovations
Innovation process	Applies the traditional stage-gate model enriched with the knowledge flows outside the organizational boundaries. Focus on inputs and outputs	Multifaceted; characterized by rapid application and "after innovation." Focus on the process
Actors involved	Mainly businesses and commercial markets	Private, public, and third sector organizations, individual citizens, and their communities
IP Management	Strong IP protection enabling patents, licensing, technology-based acquisitions, joint ventures, and non-equity R&D investments	Free access to knowledge, extensive publishing of knowledge

Open innovation has apparent linkages to social and system innovations: open modes stress the significance of collaboration and social relationships for effective innovation strategies. However, there are also differences, as shown in Table 5.1.

Differences in the focus of innovation/innovation outcome and in the IP management can be considered to reflect the early stages of open practices in social and system innovations, whereas the difference in the actors involved shows deeper difference in the nature of innovations. Also the difference in the innovation process can be interpreted in this way, but it may also indicate that social and system innovations are paving the way for a more versatile view on the processes in which innovations emerge.

In the private sector, an element of open innovation is the utilization of knowledge, and intellectual property rights (IPR) in particular, as a tradable asset. In the context of material goods, businesses may examine their IP portfolio and seek to sell or license out those intellectual assets that are not relevant for their core business. The abundant inside-out and outside-in knowledge flows involved highlight effective IP protection and management systems. The paradigm of open innovation has recently been applied to services, too (Chesbrough 2011). Here, the trading of IPR plays a minimal role, but the utilization of external knowledge is equally important.

In the public sector, open innovation covers specific networks, which consist of collaboration between public, private, and third sector service organizations in the field of innovation (Gallouj et al. 2013). However, the development of open innovation practices in the public sector has not been very rapid until now, and the focus has been mainly on the outside-in knowledge flows; the inside-out approach has received less interest. This situation is understandable to some extent, considering the responsibilities of governments to handle and protect confidential data (Lee et al. 2011). As regards nonprofit organizations, an increasing number of them are initiating a shift towards a new collaborative paradigm (Bommert 2010). They take advantage of the growing number of citizen networks and new types of online intermediates to enhance public value.

5   Summary and Conclusions

Since the mid-1990s, research into service innovation has rapidly accumulated. The next big challenge is how to take a step forward from the level of companies and organizations to broader levels on which today's most urgent issues need to be solved. It means that service innovations have to be studied and developed hand in hand with social and system innovations.

Social innovations are linked to different realms of society, but their common characteristic is that they concern complex economic and social problems. Their outcomes usually arise in the form of a service innovation, but the process of social innovation comprises much more than a traditional service relationship. The sources, goals, actors, and activities of innovation are more diverse. Social innovations may emerge at the grassroots level among individual citizens; they can be produced by private, public, and third sector organizations; or they may result in fundamental changes at the societal and policy level. Top-down and bottom-up activities are both important in the stimulation and management of social innovation processes. Top-down activities are linked to changes in policies and regulations and are often necessary for the materialization of social innovations. Bottom-up grassroots activities constitute an "engine of social innovations" and are linked to user-driven approaches in innovation.

An essential characteristic that separates social innovations from market-based innovations is the central role of dissemination: social innovations call for imitation. In them, open innovation is the primary strategy, i.e., forming alliances and networks is a core task. Due to the typically local nature, the dissemination and scaling up of social innovations require specific efforts, among which strengthening of their systemic features is an important starting point. Thus, social and system innovations are interlinked. A system innovation is based on the simultaneous development of organizations, technologies, services, and multiple network relationships. In the novelty created, new ways of interacting with other actors is an important ingredient.

Three important goals can be recognized for further research in this area. First, an improved understanding about the nature of social and system innovations is needed in order to foster and support their emergence. Second, existing methods and tools should be adapted and new ones developed for the examination and management of service, social, and system innovations and for the evaluation of their impacts. Third, policy competences should be improved to harness the benefits of service, social, and system innovations. The achievement of these goals requires both theoretical analysis and empirical evidence. Modeling the social and system innovations, case studies, utilization of statistical sources, and policy analysis would all be useful for the progress of this important research area.

References

Alam I, Perry C (2002) A customer-oriented new service development process. J Serv Market 16(6):515–534

Bommert B (2010) Collaborative innovation in the public sector. Int Publ Manag Rev 11(1):3–17

Carlborg P, Kindström C, Kowalkowski C (2013) The evolution of service innovation research: a critical review and synthesis. Serv Indust J. doi:10.1080/02642069.2013.780044

Chesbrough H (2011) Open services innovation: rethinking your business to grow and compete in a new era. Wiley, New York

Edvardsson B, Gustafsson A, Kristensson P, Magnusson P, Matthing J (2006) Involving customers in new service development. Imperial College, London

Engvall M, Magnusson P, Marshall C, Olin T, Sandberg R (2001) Creative approaches to development: exploring alternatives to sequential stage-gate models, Fenix WP 2001:17. http://www.fenix.chalmers.se

Gabriel YT, Lang T (2008) New faces and new masks of today's consumer. J Consum Cult 8(3):321–340

Gallouj F, Weinstein O (1997) Innovation in services. Res Pol 26(4/5):537–556

Gallouj F, Rubalcaba L, Windrum P (2013) Public private innovation networks in services. Edward Elgar, Cheltenham/Northampton

Giddens A (1987) Social theory and modern sociology. Polity, Cambridge

Harrison D, Klein J-L, Browne PL (2010) Social innovation, social enterprise and services. In: Gallouj F, Djellal F (eds) The handbook of innovation and services. Edward Elgar, Cheltenham/Northampton, pp 197–281

Hasu M (2001) Critical transition from developers to users—activity theoretical studies of interaction and learning in the innovation process. University of Helsinki, Department of Education

Hasu M, Saari E, Mattelmäki T (2011) Bringing the employee back in—integrating user-driven and employee-driven innovation in the public sector. In: Sundbo J, Toivonen M (eds) User-based innovation in services. Edward Elgar, Cheltenham/Northampton, pp 251–278

Hill R, Capper P, Wilson K, Whatman R, Wong K (2007) Workplace learning in the New Zealand apple industry network—a new co-design method for government 'practice making'. J Workplace Learn 19(6):359–376

Iansiti M, Levien R (2004) Strategy as ecology. Harv Bus Rev 82(3):68–78

Kaasinen E, Ainasoja M, Vulli E, Paavola H, Hautala R, Lehtonen P, Reunanen E (2010) User involvement in service innovations. VTT Technical Research Centre of Finland, Research Notes 2552, Espoo

Kivisaari S, Saari E, Lehto J, Kokkinen L, Saranummi N (2013) System innovations in the making: hybrid actors and the challenge of up-scaling. Tech Anal Strat Manag 25(2):187–201

Koen P, Ajamian G, Burkart R, Clamen A, Davidson J, D'Amore R, Elkins C, Herald K, Incorvia M, Johnson A, Karol R, Seibert R, Slavejkov A, Wagner K (2001) Providing clarity and a common language to the fuzzy front end. Res Tech Manag 44(2):46–55

Kuusisto J, Vänskä J (2011) Open innovation. In: Toivonen M, Rubalcaba L, Kuusisto J, Vänskä J (eds) Service and social innovations—policy needs and potential impacts. VTT Technical Research Centre of Finland and University of Vaasa, Unpublished working paper

Kuusisto J, den Hertog P, Berghäll S, Hjelt M, Ahvenharju S, van der Aa W (2011) Service Typologies and tools for effective innovation policy. Final Report of EPISIS—European policies and instruments to support service innovation. Tekes, Helsinki

Lapointe A, Gendron C (2004) Corporate codes of conduct: the counter-intuitive effects of self-regulation. Soc Responsibility Int J 1(1–2):213–218

Lee SM, Hwang T, Choi D (2011) Open innovation in the public sector of leading countries. Manag Decis 50(1):147–162

Lundvall B-Å, Johnson B (1994) The learning economy. J Ind Stud 1(2):23–42

Moulaert FF, Martinelli F, Swyngedouw E, Gonzalez S (2005) Towards alternative model(s) of local innovation. Urban Stud 42(11):1969–1990

Nahapiet J, Ghostal S (1998) Social capital, intellectual capital, and the organizational advantage. Acad Manage Rev 23(2):242–266

Nelson RR, Winter SG (1977) In search for useful theory of innovation. Res Pol 6:36–76

Newman J, Clarke J (2009) Public, politics and power—remaking the public in public services. Sage, Thousand Oaks

Nordlund H (2009) Constructing customer understanding in front end of innovation. Acta Universitatis Tamperensis, Finland

Payne A, Storbacka K, Frow P (2008) Managing the co-creation of value. Acad Market Sci J 36(1):83–96

Rosenthal P, Peccei R (2007) The work you want, the help you need: constructing the customer in Jobcentre Plus. Organization 14(2):201–223

Rubalcaba L, Gallego J, den Hertog P (2010) The case of market and system failures in services innovation. Serv Indust J 30(4):549–566

Rubalcaba L, Windrum P, Gallouj F, Di Meglio G, Pyka A, Sundbo J, Webber M (2011) The contribution of public and private services to European growth and welfare, and the role of public-private innovation networks. Working papers of the project ServPPIN, European Commission

Sørensen E (2002) Democratic theory and network governance. Admin Theor Prax 24(4):693–720

Sørensen F, Sundbo J, Mattsson J (2013) Organisational conditions for service encounter-based innovation. Res Pol 42:1146–1456

Sundbo J (1996) The balancing of empowerment: a strategic resource based model of organizing innovation activities in service and low-tech firms. Technovation 16(8):397–409

Sundbo J (2008) Customer-based innovation of knowledge e-services—the importance of after-innovation. Int J Serv Tech Manag 9(3–4):218–233

Toivonen M (2010) Different types of innovation processes in services and their organisational implications. In: Gallouj F, Djellal F (eds) The handbook of innovation and services. Edward Elgar, Cheltenham/Northampton, pp 221–249

Tuomi I (2002) Networks of innovation—change and meaning in the age of the internet. Oxford University Press, Oxford

Vargo S, Lusch R (2004) Evolving to a new dominant logic for marketing. J Market 68(1):1–17

van der Aa W, Elfring T (2002) Realizing innovation in services. Scand J Manag 18:155–171

von Hippel (1986) Lead users: a source of novel product concepts. Manag Sci 32(7):791–805

von Hippel E (2005) Democratizing innovation. MIT, Cambridge

Wang Y, Vanhaverbeke W, Roijakkers N (2012) Exploring the impact of open innovation on national systems of innovation—a theoretical analysis. Technol Forecast Soc Change 79(3):419–428

Windrum P (2008) Innovation and entrepreneurship in public services. In: Windrum P, Koch P (eds) Innovation in public sector services—entrepreneurship, creativity and management. Edward Elgar, Cheltenham/Northampton, pp 3–20

Chapter 6
The Limitations of Logic and Science and Systemic Thinking—from the Science of Service Systems to the Art of Coexistence and Co-prosperity Systems

Takashi Maeno

Abstract In this chapter, first, using the concepts of the uncertainty principle and the wave-particle duality of light, the science of complex systems, and the self-referential nature of logic, the author points to the limitations of logic and science, namely, that logic and science only provide a simplified model of the world. Next, from this standpoint, the problem areas of service science and its potential for development are considered. Specifically, the author does not consider services simply to be an exchange of objects, acts, and money; rather, they are a complex act with an exchange of psychological satisfaction and emotions taking place in parallel with the actual exchange.

Keywords Art of coexistence and co-prosperity systems • Science and art • Science of service systems

1   Introduction

The problems faced by contemporary society—including those relating to the environment, poverty, conflicts, resources and energy, politics and society, and science and technology—are growing in scale and becoming increasingly complicated and solving them is becoming more difficult. In a time like this, there is a need not only to systematically create solutions but also to provide a systemic overview of the whole picture. In a postmodern society where the limitations of all types of things have been pointed out, there seems to be a need to reconsider the limitations of logic and science in order to see the whole picture from a macroscopic perspective. In this paper, the limitations of logic and science are systemically investigated. The author

T. Maeno (✉)
Keio University, Tokyo, Japan
e-mail: maeno@sdm.keio.ac.jp

© Springer Japan 2015
K. Kijima (ed.), *Service Systems Science*, Translational Systems Sciences 2,
DOI 10.1007/978-4-431-54267-4_6

also considers the importance of a systemic, holistic approach that goes beyond a dichotomous worldview. Further, the limitations of and the possibilities for service science are discussed.

2 The Limitations of Reductionism and Analytical Thought

Reductionism is an approach to understanding the mechanisms and meanings of a complex system by breaking down the whole into its component parts. Each of the individual parts is analyzed and from this, an attempt is made to understand the mechanisms and significance of the more complex whole.

Reductionism is related to the mutually exclusive and collectively exhaustive (MECE) principle that was developed by McKinsey & Company, which is said to be the number one consulting company in the world. In brief, the MECE approach to analyzing and organizing a problem is to divide it into subsets that are mutually exclusive (no overlaps) and collectively exhaustive (no omissions). It is a powerful intellectual approach that can be used by anyone and at anytime whenever something can be categorized into subsets. If there are omissions or overlaps in the subsets, this will have an adverse effect on the analysis, categorization, and allocation of work. Conversely, when there are no omissions and no overlaps, it becomes possible to accurately analyze, categorize, and allocate work.

For example, a categorization of human beings into the subsets of "men" and "women" would be a MECE arrangement, as there are no omissions or gaps. However, a categorization of "men" and "girls" would create a gap (as it does not include adult women), while a categorization of "men" and "people who use women-only passenger cars" would create an overlap with the "men" subset (as men with physical disabilities or elderly men may use such vehicles.) Itemization is effective when a system or thing is being arranged and explained, with the iron rule being that itemization is a form of MECE (except for cases where this is deliberately and consciously not the case). For example, the factors that determine the change in the size of the population of Japan are as follows:

- Number of births
- Number of deaths
- Number of people coming to Japan from overseas
- Number of people leaving Japan to go overseas

These four categorizes are a MECE arrangement.

However, this intellectual approach should only be used based on an understanding of the limitations of reductionism and division thinking (MECE). Specifically, at the very least, reductionism and analytical thought (MECE) can be considered to have the following three limitations:

1. The uncertainty principle and the wave-particle duality of light
2. The science of complex systems
3. The self-referential nature of logic

2.1 The Uncertainty Principle and the Wave-Particle Duality of Light

The uncertainty principle expresses that in the world of quantum mechanics, it is impossible to precisely determine both the position and the momentum of a particle at the same time. To precisely determine a particle's position it is necessary to observe it closely, but to do so in the quantum world requires the use of light with a short wavelength. As this form of light has a high level of energy, it has a significant effect on the object being measured, to the extent that it changes its momentum. In other words, the nature of the quantum world makes it impossible to use MECE to categorize position and momentum or the observer and the object being observed.

The wave-particle duality of light describes the phenomenon that a light particle will not express the characteristics of a wave except when observed as a wave and will not express the characteristics of a particle except when observed as a particle. Therefore, due to the dichotomous nature of light, in that it expresses both the characteristics of a wave and a particle, it is impossible to describe it in concrete, black-and-white terms.

These above examples indicate that fundamentally, there are things that exist in the world that cannot be explained via reductionism and analytical thought.

2.2 The Science of Complex Systems

The science of complex systems provides examples of things that cannot be explained by reductionism. According to the science of complex systems, when nonlinear interactions take place between the elements of a system, a state of chaos may be generated within which it becomes impossible to predict the system's future state. When the behavior of a system exceeds a certain critical point, the level of disorder rapidly increases and predictions become impossible. Such systems are called complex systems, and a state of disorder in which predictions are impossible is known as chaos. Chaos is one example of an emergent phenomenon. Emergence expresses a property in which the complexity of the whole exceeds the sum of the complexity of the individual parts. Through the creation of a complex organization through multiple local interactions, a system is created that could not be predicted from the behavior of its individual parts. In other words, emergence describes the creation of a system that cannot be explained by reductionism.

The explanation of chaos that is clearest and easiest to understand uses simple, nonlinear functions, such as logistic maps and double pendulums. While the author will only attempt to provide a brief explanation of it in this paper, it is recommended that readers with an interest in this topic refer to specialist texts.

A well-known method of explaining chaos is the so-called butterfly effect, which states the way that air is disturbed when a butterfly flaps its wings in Japan will determine whether the weather will be sunny or stormy in the United States a few years later. Intuitively, it seems that the minute disturbance of air caused by a butterfly flapping its wings would be wiped out by a larger, more powerful flow of air and so would not have any effect on the weather in the United States. For example, it seems likely that even if the butterfly in Japan did have some effect on the flow of air in the United States, it would only have the most limited possible effect on climate, at the very most of about the same extent as a butterfly in the United States. However, it becomes apparent after calculations that a butterfly flapping its wings in Japan can result in dramatic climatic changes in the United States. Air flow is expressed using complex nonlinear functions and the earth's atmosphere is a colossal emergent system that includes a fantastically large number of air molecules. As a result, the actions of its smallest parts can have enormous consequences for the system as a whole.

This phenomenon, of a tiny part having an extremely large effect on the whole, is a key element of chaos. Even where there is only the tiniest difference between starting values, the results will be enormously different. It requires only a few simple calculations to confirm the existence of chaos phenomenon; a difference of only 0.001 % between the starting values will produce results as divergent as a sunny or stormy day. The same applies to economic systems; the world economy is driven by the economic activities of individuals that are agglomerated into a system of seven billion people. Accordingly, what you decide to buy for your lunch today can have enormous consequences on the economic conditions of another country a few years later. Another example is the brain, which is made up of 100 billion neurons. Information on something you casually observed a few moments ago is transmitted to your brain, where it will continue to influence brain activity for decades to come, until all activity in your brain stops. Whether your destiny is to live for decades or days, different things you have observed will have different effects on your brain activity. In other words, it is impossible to predict the future in systems where the interactions between individual elements are complex and where chaos is generated. Certain phenomenon may at first glance seem simple to model and accessible to generalized predictions, but in reality it is extremely difficult to predict the future in a system where chaos may emerge.

In other words, while it might seem that if we analyze the phenomena that occur in the world in terms of constraint conditions, field equations, and unknown parameters, and if the relations between them can be understood, then the image of a phenomenon as a whole can be captured through reductionism. However, the reality is that it is extremely difficult to make predications when chaos occurs.

2.3   The Limitations of Science

In general terms, science refers to analytical methods that are objective, unambiguous, and reproducible, and an essential precondition in the scientific method is that all subjectivity is excluded. Science assumes that the observers remain on the outside of that which they are observing. They objectively observe the phenomenon they are studying and accurately record what occurs in an unambiguous manner that will not invite misinterpretation. Thanks to this process, the results obtained from the scientific method can be reproduced by other observers in different locations.

Although they might be extreme examples, as a general rule the two previous examples from the quantum world show the limits of science. For example, the uncertainty principle means that the existence of the observer can affect the phenomenon being observed and so it is impossible to separate the two. In our daily lives, we only assume that we can separate the subjective from the objective.

In addition, strictly speaking the goals of conducting research that is "unambiguous" and "reproducible" can only be approximately achieved. In the real world, the results of modeling based on assumptions ought to be unambiguous and reproducible within the range of the model. But strictly speaking, nothing in the real world can be perfectly reproduced. Time's arrow only goes in one direction; therefore, a phenomenon that occurred at a certain time and place cannot be recreated a second time. Even if the same molecules are arranged in the same structure in this second time and place, it will still be impossible to generate exactly the same results as were achieved in the first instance. Reproducible results can be achieved in thought experiments, not in actual ones, and the only time that results can be reproduced is when simplified modeling is carried out. But the real world is full of complex systems and results cannot be reproduced excluding this sort of simplistic modeling.

Consider the behavior of a ball rolling down a slope. While it cannot be precisely reproduced at the molecular level, the movement of a ball that does not skid on a solid body can be reproduced as an approximate expression quite easily and with a reasonably high degree of accuracy by using Newtonian motion equations. While it cannot be completely reproduced, under general conditions it is possible to reproduce a rough approximation of the original event.

The scientific method aims to be objective, unambiguous, and reproducible, but in the real world these are not things that can be achieved and only in special cases can something that approximates to the original event be realized. To summarize, the scientific method is not a panacea for all problems; rather, it is nothing more than a method for verifying simplified theories in a pseudo-world.

My intent is not to deny the value of science, rather to stress that the scientific method should only be used once it is understood that it is not a panacea capable of solving all problems, but is actually nothing more than a method of verifying theories based on certain assumptions (modeling). Modern people (more accurately, modern people that tend to only understand the modern world) are prone to dismissing everything that is not science as nonscientific; but science is not something that we should place all our faith in.

The universe is entirely made up of reactions between various elements that interact in an intricate and complex way. Science entails nothing more than taking one part of the parallel, dispersive world of complex networks, simplifying it, using a series model in the form of language to ask a logic question in some way, and then providing the answer. In other words, science provides the answers to individual questions that are created out of the parts of the disassembled whole, but it does not provide answers to fundamental questions that relate to the whole.

Next, the scope of this discussion will be broadened from science to academic learning as a whole, and we will consider whether it should be objective.

Subjective learning takes place in various academic disciplines, such as qualitative research in the fields of sociology and psychology, clinical medicine in medicine, clinical psychology in psychology, business schools in economics, and management of technology (MOT) in scientific technology. None of these types of learning involve the objective, third-party evaluation of universal systems; rather, a feature they share in common is that through such techniques as dialogues and case studies, they seek individual phenomenon to research in the form of problems faced by individual people.

What is interesting is that these subjective studies are taking place in each of these academic disciplines with objective studies on the opposing axis. However, when comparing the number of researchers in each area, the overall trend is that there are fewer scholars in the subjective fields.

For example, the goal of academics in mainstream sociology and psychology is to objectively quantify the state of people in society. However, the individual characteristics of each of the separate events that are studied may be lost when statistically averaging a large number of events. The danger of averaging is that it disregards diversity and individuality. Qualitative research objects to this and stresses that it is precisely because researchers become part of the society that they are studying that they can acquire qualitative data.

The goal of economics is to systemize learning in this field to objectively elucidate economic mechanisms. But in contrast, the goal of MBA courses at business schools is to develop professionals with the management skills required to work in a complex economic system. There are also many vocational schools that provide courses other than the MBAs and none of them aim to develop students through objective, third-party learning, but rather for the students themselves to improve their own skills through first-person study. In contrast to the objective, uniform, and regression-to-the-mean approach of conventional learning that aims to construct the sole and absolutely universal method of study, each of the above courses actively recognizes subjective, diverse, and individualistic learning.

Accordingly, science and other academic fields fundamentally should not be MECE arranged into subjective and objective categories, but for the sake of expediency they can be categorized as such. Looking at this from another perspective, it can be considered that reductionism, which divides learning into its parts, tends to overlook those parts that cannot be divided.

For example, MBA students studying management methods at a business school will learn the skills to manage a company at the same time as achieving self-actualization, in the sense that they will do the work that they want to do. In fact, a third-person (objective) method of studying management will simultaneously become a means of first-person (subjective) self-actualization. One of the problems of a solely objective management approach is that discussions about subjective motivation tend to be overlooked. Of course, this is not limited to students with a first-person approach to studying; all students are third-person observers at the same time as being first person-individuals.

It has been said that since the distant past, the Japanese way of thinking has assumed that the individual cannot be separated from the whole. Japanese people have been criticized for the fact that they do not clearly separate themselves from others and establish their own self-identity. But it should be said that the fundamental activities of people cannot be divided into the self and the other, or the objective and the subjective. Even though originally the people of the world were indivisible, modern scholarship has developed from a starting point of forcibly separating people into MECE categories.

2.4   The Self-Referential Nature of Logic

Up to this point, our discussions have encompassed how the fundamental scientific approach that involves separating the self from the other and subjectivity from objectivity cannot provide explanations in the scientific domains of quantum mechanics and complex systems. This is the paradox of modern science.

The paradox of science expresses the limitations of science itself; therefore, it is not possible to resolve it from within science and instead the task of dealing with this sort of problem is fundamentally the domain of philosophy.

One of the central themes of logic is expressed in Cretan's paradox. A Cretan man says that "all the Cretans are liars." But the man himself is a Cretan and therefore, he is a liar. But if he is a liar, what he says is untrue and Cretans are truthful. But the man is a Cretan and therefore, what he says is true; but by saying that Cretans are liars, the man himself is a liar, and therefore, what he says is untrue. This is known as a self-referential paradox and cannot be solved by logic.

It is similar in form to paradoxes in science, which is as we should expect. Logic and the statements that express it are just like science in that they are nothing more than models of the real world and fundamentally speaking, the real world is a structure that cannot be comprehended through reductionism. Therefore, it should be no surprise that paradoxes are generated in both science and logic.

It is also analogous to the epistemological and ontological paradoxes that exist in philosophy. Below are some examples of paradoxes that relate to the fundamental nature of existence and knowledge.

As is expressed by Descartes in "I think, therefore I am," our awareness of the quality of self (qualia; a vivid awareness of subjective, first-person experiences) exists within us today more than ever before. However, neurological research is discovering a constant stream of evidence showing how this awareness is nothing more than a false perception or illusion. It is also considered that we do not have qualia awareness while we are sleeping or in the womb; that is to say, that awareness, while not dependent on an awareness of its own existence, is something that we can vividly experience to the extent that we cannot "not" experience it. In other words, we can say that there is no answer to the question of which is correct, materialistic monism or psychological monism. The debate about whether we have a mind because we have a brain or we have a concept of the brain because we have a mind falls into the same pattern as the question of which came first, the chicken or the egg.

The above discussion can be summarized as follows. Fundamentally, science and logic are types of reductionism. Essentially, the world is a place of nonlinear interactions that cannot be understood by dividing it into such categories as the self and the other, objective and subjective, cause and effect, reality and illusion, true and false, and logical and illogical. Categorization causes us to overlook the connections between things, but we cannot analyze anything if we take a solely holistic view and do not divide the whole into its parts. As a consequence, in modern times, we have come to adopt an approach to science and logic, as types of analytical thought or modeling thought, which divides and analyzes items that are essentially considered to be indivisible, such as the self and the other, subjectivity and objectivity, and cause and effect. We have become completely accustomed to this approach and so mistakenly consider it to be "correct," but this approach is not correct; it is merely a method of modeling the world.

The mistakes that people tend to make when modeling the world include undervaluing an objective approach, disregarding all the things that exist between subjectivity and objectivity and their interactions, and undervaluing the diverse ways of thinking exhibited by complex human beings that cannot be considered in logical terms. These types of problems tend to occur in academic learning, and the discussion below provides examples from service science to illustrate them.

3 The Problem Areas of Service Science and Its Potential for Development

Based on the context described above, we shall now consider the limitations of service science. As the limitations of science have already been discussed, we shall first look at the limitations of the conventional understanding of the term "a service." A service can be defined as a system that provides something that the public or customer needs, organized by the government or a private company. The word comes from the verb "to serve" and originally it described the service provided to a customer. In other words, service as a word expresses a clear dichotomy between the person providing the service and the person benefiting from it. However, it is possible to go beyond this dichotomy if we consider services as a system. For example, dividing a service into the person providing it and the person receiving it fails to express the subjective but essential aspect of it that exists in the mind, reducing it merely to a transaction involving an exchange of things or actions for money. In other words, we should not consider a service as simply an objective transaction, but instead it should be understood as a transaction that includes the subjective motivations of the person providing the service.

To think about this in another way, we need to consider more completely the positions taken by the person providing the service and the person receiving the service. For example, imagine that you are providing a service to a customer; what is your motivation? Is it to receive money in return? Is it to satisfy the customer? Is it to feel a sense of satisfaction from doing a good job? Each is correct. In addition to earning money, you are likely to want the customer to feel satisfied and at the same time to feel satisfied yourself with the job you have done. In other words, a service should be analyzed as an event that exists in the minds of the stakeholders that goes beyond the boundaries of a mere business transaction and beyond the categorization of the objective and the subjective. We could consider a counterargument that the way a person thinks should be consistent, but I think that this is an approach that follows too faithfully to the focus on dichotomies that has emerged in modern times. As was previously stated, the dichotomous way of thinking that has become prevalent in the modern age is fundamentally just one way of modeling the world, and I believe that the notion that human thought is consistent is nothing more than an illusion. Western people sometimes say that the Japanese are hard to understand because of the disparity between their internalized true feelings (*honne*) and the polite face that they show to the world (*tatemae*); but for me, this is only natural. Every person has and exhibits a variety of *tatemae* and *honne* and a fundamental model of the human mind is one in which decision making realizes a balance between the complex elements of the mind. I think that the Western model of the mind simplifies this to an extreme extent.

This is connected to the paradoxes that were previously mentioned; namely, is only one of *honne* and *tatemae* true, or are they both true? The answer is that the antithetical concepts of A and B are both true; this is the nature of the mind, which cannot be explained by logic. This idea is similar to the logic of *sokuhi* (the superrational logic of *is* and *not*) in Buddhism that states that "a mountain is not a mountain, therefore, it is a mountain." This seems a logical contradiction, but in Oriental-type logic it is not perceived as such. The very act of categorizing a mountain as "a mountain" or "not a mountain" is no more than a type of world modeling. However, we carry out modeling because we cannot begin to understand the world until we start to do so and so we can name a thing "a mountain." Each type of logical paradox can be resolved when we consider them in this sort of holistic manner.

Why is an understanding of an undivided world preferable to an understanding of a dichotomous service? A useful approach when thinking about this is to consider the abolition of conflict. Conflicts occur when we divide the world into "us" and "them," the enemy. When we consider that essentially, there is no division between "us" and "them," the need for conflict disappears. "Us" and "them" are simply parts of the single system of "we." At first glance, this appears to be raising the viewpoint by a meta-level; however, hierarchizing the viewpoint is in itself a form of divisionism. Rather than raising the level of the viewpoint, the viewpoint is fundamentally indivisible in nature, as the viewpoints of "I" and "we" coexist. I am I, and at the same time, I am we.

The dichotomy that exists between self-interest and altruism should also be transcended. A contradiction occurs when we consider whether the provision of services is a self-interested or altruistic act. One perspective might be that services are a business act and that the value of the service provided is the money the provider receives. But a different perspective is that the service is not provided for money, but for the customer. The former perceives services as a self-interested action and the latter as an altruistic one. But fundamentally a service cannot be separated in this way, and these two perspectives coexist in people's minds. However, if we raise the meta-level, then the service provided and received becomes simply different parts of the same system of "we." If we aspire to achieve well-being for all, then fundamentally we cannot divide self-interest and altruism. In other words, not in terms of logic, but intuitively, the fundamental goal of the human race should be understood to be peace and well-being for all. This is what we should all aim to achieve. I think that in the same way, service science should also aspire to an approach that goes beyond the boundaries of science and logic.

As was described above, the name "service science" is doubly insufficient in that both the concepts of "service" and "science" trivialize the issue. We need to move from service to coexistence and co-prosperity and from science alone to science and art. In other words, we need to leap forward from "service science" to the "art of coexistence and co-prosperity."

4 Conclusion

In this paper, first, using the concepts of the uncertainty principle and the wave-particle duality of light, the science of complex systems, and the self-referential nature of logic, the author discussed the limitations of logic and science; namely, it was asserted that logic and science only provide a simplified model of the world. Next, from this standpoint, the problem areas of service science and its potential for development were considered. Specifically, the author does not consider services to simply be an exchange of objects, acts, and money; rather, they are a complex act with an exchange of psychological satisfaction and emotions taking place in parallel with the actual exchange.

Acknowledgments The author wishes to express his gratitude to the Center for Education and Research of Symbiotic, Safe and Secure System Design, part of the Ministry of Education, Culture, Sports, Science and Technology's Global COE Program, which funded part of this research.

References

Kohtake N, Maeno T, Nishimura H, Ohkami Y (2010) Graduate education for multi-disciplinary system design and management -developing leaders of large-scale complex system design and management. Synthesiology, English Edition 3(2):124–139
Maeno T (2010) How to cultivate the faculty of reason. Kadokawa (Japanese)

Part II
Service Systems Practice

Chapter 7
Canadian Governments Reference Models

Roy Wiseman

Abstract For over 20 years, Canadian governments at all levels have been using a common set of reference models to describe the business of government. After significant experience at all levels of government in Canada, these models have stood the test of time, having proved their usefulness in many jurisdictions, consistently providing greater insight than other models less attuned to the unique characteristics of government.

These models describe the business of government from the outside-in, in terms of the programs and services that governments provide and how these contribute to achieving defined policy outcomes. This can be contrasted with the inside-out view, which focuses on how governments are organized and the activities that they undertake.

Well-constructed reference models, consisting of a common framework and language to describe the business of government, can assist in "doing government better." Focusing on outcomes and how governments are achieving those outcomes through their programs and services supports asking first whether governments are "doing the right things"—and then whether they are "doing things right."

Keywords Business architecture • Business model • Reference model

1 Introduction

Organization for the Advancement of Structured Information Standards (OASIS) defines a Reference Model as "an abstract framework for understanding significant relationships among the entities of some environment." Wikipedia defines a business model as "the rationale of how an organization creates, delivers, and captures value (economic, social, or other forms of value)." These definitions suggest that a business model can be a specific type of reference model.

R. Wiseman (✉)
Municipal Information Systems Association of Canada/Association des Systèmes d'Information Municipale du Canada (MISA/ASIM), Toronto, Canada
e-mail: roy.wiseman@outlook.com

© Springer Japan 2015
K. Kijima (ed.), *Service Systems Science*, Translational Systems Sciences 2,
DOI 10.1007/978-4-431-54267-4_7

Regarding models, George Box famously asserted that "essentially all models are wrong, but some are useful." Particularly when we move away from the domain of hard science, abstract frameworks can provide only an approximation of a reality, which often defies neat categorization. The same reality may be described using many different frameworks, all of which may be more or less appropriate—but as noted by Box, some may be more useful than others.

For over 20 years, a number of Canadian governments at all levels (federal, provincial, municipal) have been using a specific set of reference models to describe the business of government. While minor variations have evolved among the various jurisdictions, these models share a common heritage and have remained substantially the same in their core concepts. After significant experience at all levels of government in Canada, they can be said to have stood the test of time, having proved their usefulness in many jurisdictions, consistently providing greater insight than other models, many of which have their origins in the private sector and are not particularly attuned to the unique characteristics of government.

Much of the power of these models derives from describing the business of government from an outside-in perspective, in terms of the programs and services that governments provide and how these contribute to achieving defined policy outcomes. This can be contrasted with an inside-out view, which focuses on how governments are organized and the activities that they undertake.

Well-constructed reference models, consisting of a common framework and language to describe the business of government, can assist in "doing government better." Focusing on what governments do (activities and processes) leads to doing more of the same—perhaps, more efficiently. Focusing instead on outcomes, and how governments are achieving those outcomes through their programs and services, moves the discussion to a new level—asking first whether governments are "doing the right things," before any discussions of efficiency: "doing things right."

2 Canadian Governments Reference Models: A Brief History

The reference models in use by many Canadian governments have their origins in work undertaken in the early 1990s on behalf of a group of municipalities in the province of Ontario. Reflecting perhaps the first generation of business/information architecture, a few Ontario cities had undertaken to create a complete information architecture for their municipality. Understanding that this information architecture was likely to be common across municipalities, they decided to pool their resources. Under the sponsorship of the Municipal Information Systems Association, Ontario Chapter (MISA Ontario), funds were raised from 17 Ontario municipalities to engage consultants to undertake the required work.

As implied by MISA sponsorship, the work was initially IT-focused. The original proposal defined the end product as "a generic municipal model which could ultimately be used as foundation for a complete and integrated municipal information system."

This was further defined in the request for proposal (RFP), issued in January 1992 (MISA (Municipal Information Systems Association - MISA/ASIM Canada) 1992):

Systems developers are under substantial pressure to develop more integrated suites of applications. Traditionally, both developers (including in-house developers) and vendors of municipal software have been application oriented. Their efforts are seriously hampered by the lack of an overall picture of municipal data and functions. To be effective, their work must fit properly with that of those vendors/developers who are working on other pieces of the whole. Just as contractors and sub-contractors on a major construction project follow an overall set of blueprints, so systems developers need a blueprint (framework for development)—the Municipal Model.

The RFP identified two distinct outputs, which would be created by synthesizing and building on the work already done in five of the cities:

- High-level functional model (business model) defining what is done (not how it is done) in a generic municipality
- High-level data model, supported through an entity-relationship diagram, defining the data entities, attributes, and relationships needed to support the municipal business

The contract to develop what became the Municipal Reference Model (MRM) was awarded to Chartwell IRM, a small Ontario consulting firm with significant knowledge of both business architecture and local government. The subsequent product, produced by mid-1993, includes virtually all of the components of the models in use today across various governments in Canada. By agreement, the final product focused primarily on the first of the two identified deliverables—i.e., the functional model.

For the next 20 years, Chartwell (acquired in 2009 by KPMG) continued to develop and refine the basic concepts of the MRM through various consulting assignments with municipalities across Canada, as well as internationally. Towards the end of the 1990s, Chartwell was engaged by the province of Ontario to provide a similar government reference model at a provincial level—and then by the Government of Canada to undertake similar work for the national government.

The resulting products became known as the Public Service Reference Model (PSRM) for Ontario and the Government of Canada Strategic Reference Model (GSRM) for the national government. The GSRM, in turn, gave rise to the Business Transformation Enablement Program, BTEP, which was essentially a methodology for applying the GSRM for government transformation.

This work with the Canadian national government was documented in a 2005 Gartner Research Publication #G00128065 entitled *The Canadian Governments Style of Enterprise Business Architecture* (Gartner Inc 2005). This publication defines the benefits for government in applying these models as follows:

BTEP provides a methodology and a set of reference models that can be applied to help transform business incrementally. Work can proceed across different themes or communities (such as health, seniors or licensing) so that the efforts are additive. Specifically, BTEP helps government executives and project managers speak a common language and use common business models to create a common vision for the desired future state.... Furthermore, it works best when it holistically examines an area of government services across all government levels to explore constituent-centric service improvements.

The same Gartner publication references the US Federal Enterprise Architecture (FEA), as well as other similar programs internationally. It concludes that "none go as far as the Canadian program in breadth and depth."

3 Why We Need Governments Reference Models

The Gartner publication referenced above suggests that "with a bit of language substitution and with different business drivers as motivation, the essence of these modeling techniques may be applied to private sector businesses." However, one can equally argue that it is exactly the differences between the public and private sector which form the heart of these models—and which underscore the need for different models from those created for private sector organizations.

The differences between the two sectors can be illustrated by contrasting strategic plans or business strategies of public and private sector organizations. In their *Corporate Strategy for the New Millennium* (IBM Corporation 2002), IBM defines strategy as "what a company does to sustain and grow its business value into the future." The same document defines a strategic vision as "a picture in words and numbers of what the business will be at a certain point in time in the future. It includes a measurable summary financial target which, when attained, assures strategic success by generating sufficient economic value for the company to remain a desirable business entity."

This type of vision and strategy for private corporations is reflected in examples like the following, gleaned from the web sites of various private sector companies:

- "[We] will become one of the world's leading providers of inputs for plant growth by creating value for each of our stakeholders."
- "We adhere to a consistent low-risk strategy for strengthening our gold mining business and creating per share value."
- "Our vision is to be the leading energy delivery company in North America. We deliver energy and we deliver value to shareholders."

Such vision statements are frequently coupled with objectives such as these from a leading Canadian retailer: (1) strengthen core retail; (2) align all business units to reinforce the core; (3) build a high-performing organization; and (4) create new platforms for growth.

In comparison, a typical government strategic plan, such as that for the Peel Region (Ontario, Canada), includes the following vision (Region of Peel 2011): "Peel is a safe, healthy, prosperous, sustainable and inclusive community that protects its quality of life." Peel's Strategic Plan identifies seven overall goals: (1) protect, enhance, and restore the environment; (2) build a community that is stable, responsive, and adaptable; (3) maintain and improve the health of Peel's community; (4) support and influence sustainable transportation systems; (5) build a cohesive Peel community; (6) ensure a safe Peel community; and (7) strive for continued excellence as a municipal government.

As can be seen, private sector strategic plans are internally focused, based on strengthening the capacity and position of the corporation and increasing shareholder value. In contrast, public sector strategic plans are focused externally on addressing community needs. The goals in private sector strategic plans are expressed in terms of outcomes for the company—the provider of the goods and services. For the public sector, goals are expressed in terms of outcomes for the citizens—the consumers of the goods and services.

Even if there were no other differences between government and private sector organizations, these fundamental differences in business motivation would argue for a different business model.

4 Canadian Governments Reference Models Essential Concepts

The following describes some of the key concepts in the Canadian Governments Reference Models (CGRM). While it is not possible to fully explore all features of these models in this summary, further documentation is available from the references cited at the end.

The key features of the CGRM are illustrated in Fig. 7.1—a diagram which has become familiar to many Canadian governments at all levels.

This diagram suggests that governments can be understood in terms of the following concepts:

- The programs that they provide that address the needs of their constituents (citizens, clients)
- The specific services which deliver outputs to clients and contribute to program outcomes

Fig. 7.1 Canadian governments reference model (CGRM) key components

- The processes which are part of delivering those services
- The resources used in carrying out those processes

As noted previously, this description of government provides an outside-in view, based on the impact that governments have on their constituents or citizens. In contrast, an inside-out view would focus on the activities of government, how they are organized and the resources used.

The left side of the diagram defines the governance relationships between the provider organizations (essentially government departments/organization units) and the programs, services, and processes. Missing from this diagram is the broader governance relationship between citizens and their governments. While governments and their departments provide programs and services to their citizens, it is the citizens, through the democratic process, who should determine which programs and services are provided and at what level, including their relative priorities and allocation of resources.

4.1 Structure of the Model

Before describing some of the components illustrated in Fig. 7.1, it is worth noting that the overall meta-model includes:

- A number of *components* or *entities* (e.g., programs, services, outputs), each with a precise definition and set of rules for determining valid *instances*, *properties*, and *relationships*.
- A set of *properties* or *attributes* which may or must be provided for each component and which further define and describe the component. For instance, each service output has a property of service output type.
- A set of *relationships* which may or must be declared among the components. These may take various forms: one-to-one, one-to-many, many-to-one, etc. For instance, a service is administered by one program, but one service may contribute to the outcomes of a number of programs.
- A number of *value lists* which provide open or closed sets of possible values which can be assigned to a given property. For instance, each service output has a service output type, selected from a closed list of nineteen possible service output types.

In addition to the meta-model, which is common across all levels of government, the municipal (MRM) version provides pre-populated instances for many entities and properties, including:

- A set of municipal programs and several hundred municipal services, drawn from actual program and service inventories compiled by individual municipalities across Canada—validated to ensure that they conform to the model rules and definitions. Both the programs and services are arranged in hierarchies of program–subprogram and service–subservice.

- Each program or service has a pre-populated set of attributes and relationships. For instance, each service has a service description, service output, service output type, and one or more program outcomes to which it may contribute.
- Some attributes and relationships are left unpopulated, where these must be determined uniquely be each municipality applying the model. For instance, a municipality may populate the model with its own organizational structure and assign responsibility for administering a program or delivering a service to the appropriate organizational unit.

This pre-populated program and service catalogue is not intended as a standard defining a complete set of the "right" set of programs and services with the "right" properties and relationships. Rather the purpose of the catalogue is to provide both a quick start to municipalities wishing to apply the model and a concrete illustration of the principles underlying the model. Experience has shown that jurisdictions will gain value from the model, only by going through the effort of understanding those principles and applying them to developing their own service catalogue. Like many other such methodologies, the process is as important as the product.

4.2 Key Concepts: Government Program

Key to the Canadian Governments Reference Models is the concept of a government program. While this term is used widely in governments, often in a budgeting context (funds are allocated to programs), the term is rarely defined and may be applied to a broad range of government activities (services, activities, projects) from the very large (e.g., Public Safety Program) to something quite specific (Drinking Water Surveillance Program).

In contrast, the CGRM provides the following precise definition of a program:

Program: a mandate conferred from the governors of the enterprise to achieve outcomes that address identified needs of a target group.

This definition incorporates the following concepts:

- Programs derive from a mandate provided from "the governors"—i.e., elected representatives and, ultimately, from the electors.
- Programs are intended to achieve *outcomes*, which must be defined.
- Outcomes, in turn, meet identified *needs* of defined *target groups*.

Not included in the definition, but implied by the associated rules, properties, and relationships in the model, are the following:

- Programs are delivered through a set of services.
- Programs, through their constituent services and processes, consume resources.
- Programs may be further classified into subprograms, based on a further refinement of the target group or need. For instance, the Public Safety Program includes subprograms such as Safety from Crime, Fire Safety, Construction Safety, and Transportation Safety.

- Programs provide a basis for business or service planning to determine the optimal mix of services required to achieve the identified program outcomes.
- Programs are often used as a management and budgetary construct, providing an accountability framework and a basis for funding allocations.
- While a program represents the intersection of a set of *needs* and a *target group*, the names for many government programs reflect just one of these concepts. For example, a housing program references just the need, whereas a seniors program references the target group. The implication may be that the housing program is intended to address housing for *all* residents or that the seniors program will address *all* needs of seniors. However, in analyzing such programs, it is worth looking more closely at both the need and the target group to see if each can be more precisely defined. For instance, the target group for the housing program may be just low-income residents.

For governments, programs are fundamental instruments of public policy. They define the needs which the government is addressing or intends to address.

The private sector equivalent to a government program may be a line of business; however, the rationale for a line of business is essentially whether it can generate a greater profit (shareholder value) for the corporation than alternative uses of the same funds. Addressing a need is a means to an end, rather than an end in itself. Asking whether a proposed line of business is addressing a consumer need can be understood as shorthand for whether there is a market for the proposed product or service.

In contrast, a new government program starts with the need—and specifically, a need that is within the government's mandate to address and which the government, with input from its electors, has decided to address. If services (see below) describe "what" a government does, programs provide the "why."

4.3 Key Concepts: Government Service

As noted above, governments achieve program outcomes through delivering a set of services. Rather than directly delivering a unit of public health (a program), a government delivers a set of services, each of which is intended to contribute to the defined public health outcome. For the citizens and clients of government, programs are abstract, whereas services are real. They represent the point of interaction between governments and their clients.

Like program, the term service is used widely in both government and the private sector with a range of definitions. The CGRM provides the following definition:

Service: provision of specific final outputs that satisfy client needs and contribute to program outcomes.

This definition incorporates the following concepts:

- A service must have a *client*. Internal processes (like service planning) are not services, since there is no client.
- Services transfer a valued *final output* to that client—for instance, a unit of swimming instruction (swimming lesson). On the other hand, registration for the

swimming instruction is not a service. The final output sought by and delivered to the client is the swimming instruction. The registration is an intermediate output of the registration process.

- Services contribute to *program outcomes*. For instance, an immunization service is intended to reduce the incidence or severity of the disease for which the client is being immunized. This, in turn, contributes to an overall public health program outcome.

Not included in the definition, but implied by the associated rules, properties, and relationships in the model, are the following:

- Services may contribute to outcomes for more than one program. For instance, a Solid Waste collection service may contribute to outcomes related to public health, environmental protection, and economic development.
- The *service value* defines the relationship between a service and the programs to which it contributes. For instance, service values from residential waste collection might include:
  - Reduced disease, associated with uncollected waste (contributes to public health outcome)
  - Reduced environmental impact from diversion (recycling, reuse) of collected waste (contributes to environmental protection outcome)
  - Protection of property values, due to avoidance of uncollected waste (contributes to economic development outcome)
- Each service produces only one primary *service output*.
- The service output has a *service output type*, selected from a closed list of 19 output types. These service output types provide a useful way of classifying services. For instance, the period of permission output type identifies all licensing-type services; the funds output type identifies all services where the output is funds (amount of money) provided to the client.
- Services may be further classified into sub-subservices, based on further refining the description of the client or output. For instance, a recreation instruction service may include subservices for different types of instruction (swimming, skating, dance, etc.). Similarly, a swimming instruction service may be further classified by client group: swimming instruction for adults, for children, for toddlers, etc.
- Services are made up of *processes* (e.g., registration) which deliver intermediate outputs (i.e., to other processes).
- Services, through their constituent processes, consume *resources*.
- A service cannot depend essentially on the existence of another service. If it does, it is likely a process of that other service. For instance, permit or license compliance monitoring is a process of the corresponding permit or licensing service, since it would not exist without that service. (The monitoring "service" also fails the test of providing a final valued output to a client.)
- A service may have direct and indirect clients. *Direct clients* receive the service output; *indirect clients* benefit from the service, but do not directly receive the output. For instance, the direct client for the taxi licensing service is the taxi owner/operator, who receives the license (permission to operate). The taxi patron/user is the indirect client and receives the presumed benefits (enhanced safety, improved service quality) from using a licensed operator. The presumed benefits

(service value) to indirect clients often provide the rationale for the service (link to program outcomes). Many government services (like taxi licensing) are undertaken primarily for their benefits to the indirect clients (taxi patron).

- The client of a service must be external to the organization providing the service. Internal activities, like service planning or budgeting, are not services. More precisely, they are not services provided by government to its citizens. They may, however, become services of a special class, called *internal/enabling services*, when provided by one government unit to other government units—i.e., internally centralized or shared services.

The CGRM definition does not make the traditional distinction between goods and services—perhaps because almost all of what governments provide fits the traditional definition of a service—e.g., "a type of economic activity that is intangible, is not stored, does not result in ownership, and is consumed at the point of sale." Even where they provide what might be defined as "goods" (e.g., drinking water, funds), governments tend to "service-ize" the output. Rather than simply a unit of water or an amount of funds, governments describe their outputs as a "safe, assured supply of water" or "financial assistance." This same tendency to "service-ize" goods is found in private sector marketing, which focuses on the "customer experience," rather than the mere delivery of a commodity. This modern tendency to call everything a service is reflected in yet another recent definition of a service: "acts performed for others, including the provision of resources that others will use."

In any case, further partitioning government offerings into goods and services would add little value to the CGRM models. The 19 service output types can be used to classify services, in terms of whether they produce tangible or intangible outputs. Some output types (units of resource, funds) correspond most closely to "goods," while most (e.g., advisory encounters, educational and training encounters, periods of permission, periods of protection, etc.) correspond to traditional definitions of a service.

Fundamental to the CGRM definition of a (government) service is the focus on contribution to program outcomes. Rather than being delivered of and for themselves (or to generate revenue), government services must be understood in terms of the program outcomes to which they contribute and which provide both their rationale for being and the basis for evaluating their effectiveness.

4.4 Additional Concepts

As noted previously, it is not possible in this summary to fully explore all of the concepts included in the CGRM. Fortunately, some concepts, like resources, are used in ways consistent with their use in other (private sector) business models. However, the following points may assist in further describing the overall model:

- Some Canadian jurisdictions have created taxonomies for both needs and target groups, although no standard taxonomy has been adopted. Such taxonomies can be useful for various purposes, including service bundling—presenting related services together in the expectation that an individual needing one service may also need a related service.

- Understanding the intended outcomes, needs, target groups, service outputs, and service values leads directly to performance measures (effectiveness, efficiency, and quality) for a program or service. For instance, measures are frequently derived as follows:

  - Program effectiveness—a measure, either directly or by proxy, of whether the outcome is being achieved
  - Program efficiency—cost per target group member
  - Service effectiveness—a measure, either directly or by proxy, of whether the defined service value(s), contribution to program outcomes, is being achieved
  - Service efficiency—cost per unit of output or cost per client (service recipient)
  - Service quality—conformance with an established standard or documented process, sometimes measured by client satisfaction with the service
  - Note that these measures are independent of each other. One can have a high level of satisfaction with a service that is quite ineffective—and vice versa. For instance, satisfaction (or otherwise) with how an immunization is delivered will depend on factors such as wait times and staff courtesy—which may have little correlation with whether the immunization is effective in preventing the disease.

- The CGRM also defines a number of reference patterns, which can be used to generate instances of a given entity. For instance:

  - Template performance metrics have been defined for each service type (related to service output types). A service of type "provide funds" might have different performance metrics that a service of type "provide care and rehabilitation encounter." On the other hand, different services within the same service type might be expected to have similar metrics.
  - Similarly, a template set of processes have been defined for each service type, under four main headings: service planning; service provisioning; service delivery; service decommissioning. Again, different services of the same service type can be expected to have similar processes, although the detailed steps making up these processes will vary.

- The CGRM includes a number of diagrams which illustrate the relationships among the components. For instance:

  - Program logic model (PLM) aligns direct, intermediate, and strategic outcomes with service outputs showing the chain of value from a specific service (service value) to subprogram and program outcomes. A sample program logic model is provided as Fig. 7.2 below.
  - Program-service alignment model (PSAM) aligns target groups and recognized needs with service outputs and programs (vertical accountability). A PSAM can be used to identify overlaps or gaps among programs, in terms of their target groups or needs.
  - Service Integration and Accountability Model (SIAM) depicts accountability relationships between organization units and the services within a program—or the processes within a service.

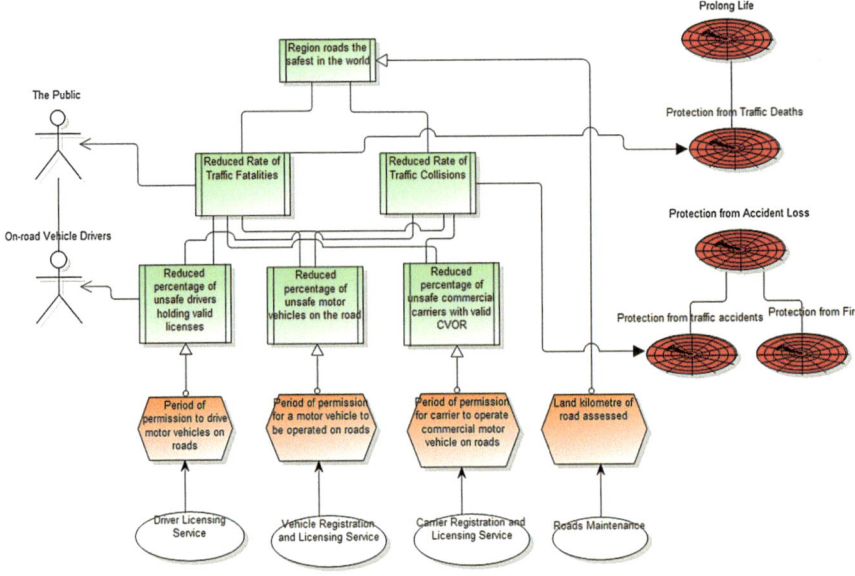

Fig. 7.2 Sample program logic model (PLM)

4.5 Putting It All Together

When the elements in the CGRM are put together, they yield a powerful understanding of what government is doing—and why. For instance, Table 7.1 illustrates, in tabular form, three of the many services that may make up a public health program in a given jurisdiction.

The linkages between a program and its services can also be provided in graphic form, using a program logic model, as in Fig. 7.2, this time illustrating some of the services of a transportation program (roads subprogram).

5 Case Studies and Use Cases

The Canadian Governments Reference Models have been applied by jurisdictions across Canada and internationally to address a range of needs and opportunities. To date, more than 40 different use cases have been identified, including:

- Strategic planning—aligning strategic plans around the major programs offered by the jurisdiction and the high-level outcomes to be achieved by each
- Program and service reviews—ensuring the optimal mix of services to achieve identified program outcomes

Table 7.1 Sample program/service profiles

Program	Public health		
Target group	All residents		
Need	Health		
Outcome	Improved health		
Effectiveness measure	Reduced incidence of disease, improved health and fitness index, increased average life span		
Efficiency measure	Cost/resident (cost per target group member)		
	Service 1	Service 2	Service 3
Service	Food premise inspection	Flu immunization	Prenatal education
Service output	Inspection report	Immunization	Prenatal class
Service output type	Period of permission (to continue operating)	Period of protection	Educational and training encounter
Direct client	Food premise operator	Resident receiving immunization	Expecting parent
Indirect client	Food premise patrons	Other residents and visitors	Newborn child
Service value	Reduce illness resulting from improperly prepared food	Reduced incidence or severity of flu	Improved level of care for newborn child
Effectiveness measure	Reduced illness from improperly prepared food; reduction in violations or exceptions	Reduced incidence of flu for those immunized/or all residents; coverage achieved (% of target group immunized)	Observed improvements in care of infants; reduced negative incidents in families receiving training
Efficiency measure	Cost/inspection; cost/operator	Cost/immunization	Cost/class; cost per participant
Quality measure	Inspection carried out in conformance with standards; operator satisfaction with inspection process	Immunization delivered in conformance to standard; recipient satisfaction with process	Classes start and complete on time; attendee satisfaction with presentation and content; attendee perceived value
Processes	Schedule inspection, conduct inspection, prepare report, schedule follow-up	Provision clinic, prepare public communication, operate clinic, deliver immunization, measure and evaluate results	Schedule classes, prepare public communication, register parents for classes, deliver class, measure and evaluate results

- Organizational/governance reviews—ensuring clearer accountability for results (program outcomes) and, where appropriate, better aligning organization or governance structures to achieve those results
- Service-based budgeting—restating the budget in terms of programs and services, rather than by department, as a basis for better understanding and communicating the implications of service cuts or service level adjustments

- Performance management—defining effectiveness, efficiency, and quality measures, based on program and service outcomes, including public reporting of outcomes and results
- Information technology planning—mapping IT applications to programs, services, and processes to identify opportunities and gaps; also identifying services with similar processes (same service output type) which lend themselves to shared applications
- Information management—using programs and services as a foundation for an information management taxonomy
- Service Bundling/Service Integration—providing single or integrated access to sets of related services addressing a common need or target group

The following case studies illustrate some of these use cases.

5.1 City of Fredericton: Service Quality/ISO Certification

In 1997, the City of Fredericton (New Brunswick, Canada) was only a $65 million business with about 600 employees (Stapleton 2004). But like many municipalities, it was providing more than 150 distinct services, as well as 46 corporate initiatives and projects, and many more departmental projects. At the time, the City had 11 departments, 106 internal committees, 96 external committees, and 9 council committees. For some management staff, participating in committees was virtually a full-time job.

Paul Stapleton, City Administrator, hoped to use a strategic planning process to create structure from this apparent chaos. In researching various approaches, the city staff discovered the Municipal Reference Model and felt that it might address their need for a common language and approach for setting organizational priorities.

Over a three-month period, the staff from the city were engaged in a number of workshops to apply this model, including:

- Creating an initial list of 19 programs, 161 public services, and 115 internal services provided by the City.
- Developing descriptions, outputs, and outcomes for each of the programs and services, including aligning each service to at least one program.
- Assigning organizational leadership for or involvement in delivery of each service, including resolving duplications, overlaps, or gaps in accountability.
- Identifying efficiency and effectiveness metrics for each service.
- At each step, reviewing and refining the list of services. After the final review, the initial list was reduced to 7 programs, 112 public services, and 35 internal services.

Using the results from these workshops, the City Council approved a new organizational structure, with departments and Council Standing Committees essentially mirroring the identified programs. This new structure was then used as a basis for an updated strategic plan, IT strategic plan, and the new budget. By 2004, Fredericton became the first municipality in Canada to achieve ISO 9001 certification—the culmination of a five-year quality management journey that began with their implementation of the MRM.

Fredericton's MRM experience has been described as follows by Paul Stapleton:

> The City needed a management tool to help *turn the organization inside-out*, that is, change the organization from a group of inward-focused departments to an outward-focused, customer-oriented, quality service delivery organization. The management tool which the City chose was the Municipal Reference Model (the MRM). … It was the right tool, at the right time, in the right place. The MRM required the staff to question everything we do, how we do it and why. It allowed us to identify our seven core programs and our 150+ services. It forced us to focus on our customers and service delivery. …Using the tool promoted teamwork and understanding throughout the organization. …Perhaps most importantly, using the MRM forces you to link your services together into programs and thereby put your work into the context of a broader social good. This, in turn, opens the door to opportunities to continually improve, co-operate and take a creative approach to improving the quality of life in your community.

5.2   City of Toronto: Core Service Review

In 2011, following the election of a new mayor and under increasing fiscal pressure, the City of Toronto engaged consultants to undertake a full Core Service Review examining which services the City should be delivering, as well as reviewing service levels and how the services were being delivered to ensure the most efficient and cost-effective delivery (City of Toronto 2011).

Fortunately, the City had already been using the MRM for some time to define city services and to understand the cost of each service. This was part of an initiative to introduce service-based budgeting at the City.

Using the existing MRM work as a foundation, the consultants (KPMG) classified the already defined services as: mandatory, mandated or required by legislation; essential, critical to the operation of the City; traditional, provided by virtually all large municipalities for many years; and other, provided to respond to particular needs or specialized purposes.

They next analyzed whether the service was being provided at, below, or above standard. As part of this, the standards being used were classified using the following hierarchy:

• Required by legislation
• Consistent with industry standards and practices
• Consistent with business case analysis justification
• Consistent with service levels in other municipalities
• Consistent with reasonable expectations

Applying this framework, a standard defined by legislation would carry greater weight than one defined internally—which might be reevaluated to see if it could be reduced without negatively affecting outcomes and/or satisfaction. Similarly, performance above standard might suggest an opportunity to reduce service levels.

Finally, the consultants identified a set of opportunities associated with each service, including options to:

- Divest, transfer, discontinue, or significantly alter the service
- Achieve program outcomes with a different mix of services
- Adjust service levels or standards, where not legislatively set
- Outsource, insource, or change a procurement approach for the service
- Improve efficiencies through redesigning processes, tools, and key enablers

The first phase of the Core Service Review, identifying high-level opportunities and potential cost savings, was completed within approximately two months. This was possible only because the City had already completed the groundwork of defining its services, including outputs, outcomes, service levels, and associated costs.

5.3   Seniors Services Initiative: Service Transformation

Between November 2003 and February 2004, participants from the Government of Canada, Province of Ontario, and municipalities within Niagara Region undertook to identify transformation opportunities in the delivery of services to seniors provided by all three levels of government (Chartwell Inc 2004).

Applying the federal (GSRM) version of the CGRM, including the BTEP methodology, the work was carried out in the following steps:

- Reviewing the vision, mission, and goals of all three levels of government related to services to seniors to identify common themes.
- Further defining the relevant target groups. For instance, the target groups include not just seniors but also caregivers (family members and others in the community providing services to seniors). In addition, the definition of a senior varied, depending on the service offered.
- Defining the relevant needs for each target group. For seniors, these included shelter, health care, nutrition (meals), financial, social needs, recreational needs, etc. The needs of caregivers or caregiver organizations were similarly defined.
- Documenting the relevant services provided by each jurisdiction which addressed these needs. The result was a list of 143 public services (services provided directly to seniors) and 60 provider services (services provided to provider organizations—including another level of government—e.g., long-term care funding provided from the province to municipalities).
- Creating service profiles for all 203 services, including service description, output, client, administering program, responsible jurisdiction/organization, service output type, and target group/need being addressed.

The resulting analysis, facilitated by many of the GSRM and BTEP tools (PLM, PSAM, SIAM, etc.), identified many opportunities, duplications, and service gaps, which were prioritized into three transformation opportunities:

- *Streamlining data collection and management*: 142 different services required clients to provide basic personal information. While a given senior may access only a small number of these services, having to provide this information over and over again can be extremely frustrating. The opportunity was to create one record on a senior, containing basic, up-to-date information (name, address, date of birth, etc.), that the senior and pre-authorized service providers could access (contingent on their need to know for the types of services they provide). This information would be shared among trusted service providers—those who already have this personal information in their own separate records. In addition, changes to such basic information would only be entered once, and all service providers would have this updated information, avoiding the need for the client to repeatedly inform all services providers themselves.
- *Streamlining information provision regarding available services:* 17 different services from all three levels of government were primarily focused on providing information to seniors. A single integrated service could provide seniors with information about available services from all levels of government at once. Such information would be more complete, consistent, packaged, and delivered to seniors at the right time. This opportunity has since been implemented in many communities through a common online seniors portal, providing one stop access to information on services to seniors provided in their community by any level of government, as well as by nongovernment agencies.
- *Collaborative case management:* The analysis showed that a single senior could be the subject of up to eight different services involving case management, each addressing a single need of the senior and each with its own eligibility reviews and extensive case files. The opportunity was to coordinate complex services delivered by different service providers, providing holistic needs assessment and the ability to broker services. It would mean treating the senior holistically—doing the "way-finding" and integration work for them, rather than expecting seniors (or their families) to do it for themselves. It would avoid the senior being asked the same care-related questions by multiple case managers and health-care professionals. The service providers would also benefit from casework already undertaken, building on services already offered and any lessons learned

6   Future Directions

Notwithstanding their use over many years and in many jurisdictions, there is no longer a single source of information about these models. As a volunteer organization, MISA was unable to sustain the effort to maintain the MRM, although it retained the intellectual property ownership. Similarly, the use of these models in various national and provincial governments ebbed and flowed with changes of leadership. For many years, interest was maintained through consulting services provided by Chartwell and others, applying these tools to address different sets of needs among governments at all levels in Canada.

More recently, MISA renewed its interest in the MRM, through an MRMv2 project, and developed a governance structure (standards board) and approved set of documents. MISA has placed a number of these documents on its MISA/ASIM Canada web site at www.misa-asim.ca. This site also includes a forum for discussing MRM-related concepts and use cases. With a user base of 30–50 municipalities, each of which has had to learn to use these models in relative isolation (with some consulting support), one of the key identified needs is for users to be able to more effectively share knowledge and experience, ask questions, and receive guidance from others.

Another aspect of the MRMv2 project has been implementation of the MRM in business modeling software. Through a partnership with IBM, MISA has supplied the underlying MRM content which IBM loaded into their Rational System Architect software and is currently making available to its clients and partners. While municipalities have for many years implemented the MRM and maintained their service catalogues using nothing more sophisticated than word processing and spreadsheet tools, use of business modeling software can provide enhanced maintenance and analysis capabilities for more sophisticated users. In addition, since the arrangement with IBM is nonexclusive, MISA remains open to mutually beneficial arrangements which would support implementation of the MRM using software from other vendors.

The Province of Ontario has published some very good documents, describing its PSRM version as part of its Government of Ontario Information Technology Standards on the GO-ITS web site at http://www.mgs.gov.on.ca/en/IAndIT/STEL02_047303.html—see especially Standard 56.1, Defining Programs and Services in the OPS (Ontario Public Service) (Government of Ontario 2010). Ontario has also experimented with using business modeling software in conjunction with the PSRM.

Finally, the Government of Canada has a substantial number of GSRM documents, perhaps the largest collection of all, but has not yet made these publicly available. In addition, many documents describing the BTEP methodology for using the GSRM were once publicly available, but these have now been archived (Government of Canada, Treasury Board Secretariat 2004).

Collectively, Canadian jurisdictions are recognizing the potential loss of knowledge associated with these models and are taking steps to avoid this outcome. The Service Mapping Subcommittee (SMSC) of the Joint Councils, representing the CIO and Service Delivery communities from all three levels of government in Canada, has undertaken the work to document existing variations of the model and to recommend a common standard which could be adopted by all jurisdictions— while recognizing that some variation may be inevitable given the history and different needs of each jurisdiction. SMSC has also sponsored a number of webinars on the application of these models and made these available through the ICCS Media Library at http://www.iccs-isac.org/about/media-library/?lang=en (Institute for Citizen-Centred Service 2009).

Finally, MISA is initiating the process of proposing the MRM and/or CGRM as a standard to an appropriate standards body—e.g., the OMG Government Domain

Task Force. In this regard, considerable work has already been done, with the assistance of IBM, in mapping the MRM to both the Business Motivation Model (BMM) and The Open Group Architecture Framework (TOGAF).

7   Concluding Thoughts

While initially conceived as a data model that would assist the development or acquisition of IT applications, subsequent uses of the MRM and related models took it far away from these roots. This reflects the power of its basic concepts in providing new insight for describing (and then improving) the business of government. As evidenced by the use cases, the appeal of these models was often to senior business leaders (city manager, deputy minister, strategic planner, policy advisor) rather than to IT leaders or CIOs.

Despite this broader interest, the responsibility for the MRM remains with MISA/ASIM Canada (an IT-based association). Similarly, responsibility for implementing both the PSRM in Ontario and the GSRM at the Government of Canada is, in both cases, under an architecture program reporting to the CIO. This reflects a continued perception that "architecture," including business architecture, is an "IT thing" to be applied primarily in the context of implementing new IT systems.

In their 2005 report, cited previously, Gartner predicted that "Governments are increasingly recognizing that incorporating business architecture is essential to enabling transformation and to moving beyond IT-centric EAs." Unfortunately, this transformation is still somewhat in its infancy. Notwithstanding many examples of success, including those cited here, the use of architecture-based business models is not yet standard practice for governments, in Canada or elsewhere.

But perhaps the final word on the use and value of these models can be provided by Paul Stapleton, former City Administrator for the City of Fredericton, who described the MRM as follows (Stapleton 2004):

> **What the MRM is not:** it is not the solution to a problem nor is it a consultant's report that tells you what you should do. While the process is fast, it is definitely not easy. The model doesn't set priorities for you, although it allows you to discuss them using a common language and in reference to clients and service delivery units, which is helpful.
>
> **What the MRM is**: it is a customer-focused tool that forces the service providers to look at themselves from the outside in. It allows a relatively detailed analysis of services in a short time. It creates common language and measurement tools that can be logically applied to apparently unrelated functions. When efficiency and effectiveness measurements form the basis of the discussion, it becomes much easier to make hard decisions because you are dealing in measurable quantities. Using the tool promoted teamwork and understanding throughout the organization as a relatively large group of managers participated in the exercise.
>
> Perhaps most importantly, using the MRM forces you to link your services together into programs and thereby put your work into the context of a broader social good. This in turn opens the door to opportunities to continually improve, co-operate and take a creative approach to improving the quality of life in your community.

References

Chartwell Inc (2004) Seniors service mapping initiative—Veterans' Affairs Canada, Final Report. www.iccs-isac.org/members/smsc/docs/.../fed/seniors/seniors_report.doc

City of Toronto (2011) Service review program. http://www1.toronto.ca/wps/portal/contentonly? vgnextoid=2ba485a6820c3410VgnVCM10000071d60f89RCRD

Gartner Inc. (2005) The Canadian governments style of enterprise business architecture, Analysts: Scott Bittler R, Gregg Kreisman, Gartner Publication Number G00128065. https://www.gartner. com/doc/482356/canadian-governments-style-enterprise-business

Government of Canada, Treasury Board Secretariat (2004) Business transformation enablement program—an executive overview. http://www.collectionscanada.gc.ca/webarchives/2007112 5180244/http://www.tbs-sct.gc.ca/btep-pto/index_e.asp

Government of Ontario (2010) GO-ITS 56.1 Defining programs and services in the Ontario Public Service, Appendix A. http://www.ontario.ca/government/go-its-561-defining-programs-and-services-ontario-public-service-appendix

IBM Corporation (2002) Corporate strategy for the new Millennium, by Saul J. Berman, Peter J. S. Korsten. www-07.ibm.com/services/pdf/corp_strat.pdf

Institute for Citizen-Centred Service (2009) Canadian governments reference model. http://www. iccs-isac.org/research/cgrm/?lang=en

ICCS (Institute for Citizen-Centred Service) (2012) The municipal reference model—understanding the DNA of Government, presentation/webinar by Wiseman R, Amsden J (IBM). Available from ICCS Media Library at http://www.iccs-isac.org/about/media-library/?lang=en

MISA (Municipal Information Systems Association—MISA/ASIM Canada) (1992) Municipal model project RFP. Available from MISA/ASIM Canada, www.misa-asim.ca

Region of Peel (2011) Region of peel strategic plan 2011-2014. http://www.peelregion.ca/corpserv/ stratplan/plan.htm

Stapleton P (2004) Fredericton's quality journey. Unpublished article

Chapter 8
What Is 5S-KAIZEN? Asian-African Transnational and Translational Community of Practice in Value Co-creation of Health Services

Hiro Matsushita

Abstract This chapter examines some of the translational features of value co-creation by focusing on such managerial human activity systems as quality improvement, organizational learning, plan-do-check-act cycle, and cross-cultural diffusion of managerial systems of health services. By so doing this work descriptively analyzes some aspects of value co-creation in action research-based learning in health services administration.

Keywords Action learning • Community of practice • Quality improvement • Value co-creation

1 Introduction

The five Ss are originally derived from the Japanese words "seiri," "seiton," "seiso," "seiketsu," and "shitsuke." In English the five Ss mean "sort," "set," "shine," "standardize," and "sustain." The sequence of 5S focuses on effective workplace organization and standardized work procedures. As such the combination of 5S (sort, set, shine, standardize, and sustain) and KAIZEN (continuous improvement) originated in the operational management methodology of Japanese manufacturing sector. In recent years, however, the objects of 5S-KAIZEN have been transferred from the traditional manufacturing activities to the value co-creation activities of service sector including healthcare and medical services. 5S-KAIZEN is now utilized not only in

---

The earlier version of sections 7 and 8 was included in Change Management for Hospitals, a project document of Japan International Cooperation Agency, 2013 (Matsushita 2013).

H. Matsushita (✉)
Kanagawa University of Human Services, 1-10-1 Heiseicho, Yokosuka 238-0013, Japan
e-mail: sparklingmetal@gmail.com

© Springer Japan 2015   129
K. Kijima (ed.), *Service Systems Science*, Translational Systems Sciences 2,
DOI 10.1007/978-4-431-54267-4_8

Japanese health services but also in the global community including Asia and Africa. These days a number of health services institutions have introduced and effectively used 5S-KAIZEN in order to improve the levels of quality, safety, and work environment in African countries.

As a consultant of health services systems, the author has been involved in sharing 5S-KAIZEN method with the leading practitioners in health services in such countries as Benin, Burkina Faso, Burundi, Eritrea, Kenya, Madagascar, Malawi, Mali, Morocco, Niger, Nigeria, Senegal, Sri Lanka, Tanzania, and the People's Republic of the Congo. Based on the participatory observation and action research, the purpose of this chapter is to introduce the 5S-KAIZEN methodology from the perspectives of systems thinking and service systems management. The focus is to see and describe the methodology from a systemic point of view by reviewing the improving practices in African health services. First, this essay briefly reviews the stories of practices of the 5S-KAIZEN methodology in some of the African countries. Second, the translational aspects of the methodology will be described. Third, systemicity of the methodology will be descriptively analyzed from such perspectives as (1) holism and hierarchy, (2) communication and control, and (3) evolution.

2   Story One: Sri Lanka

Dr. Wimal Karandagoda, president of Castle Street Hospital for Women, began applying 5S-KAIZEN to improve quality of care, safety, and job satisfaction for the first time in Sri Lanka in 2000. When he was appointed the position he found that the hospital was very dirty and dilapidated; many patients were dying due to the poor level of patient care (Handa 2012). Learning the 5S-KAIZEN widely practiced in the manufacturing industry in Sri Lanka, he began using 5S to improve the work environment of his hospital. Initially the labor union of the hospital was very against his attempts; however, the employees gradually began practicing 5S.

These improvements paved the pathway for the decrease in the maternal, neonatal, and perinatal birth rates and stillbirth rates. Further, there is an improvement in the postsurgical infection rates. The preventable maternal death is zero for the past few years. In 2004 Castle Street Hospital for Women was awarded "Quality Award" by the ministry of industry of Sri Lanka. Due to the following achievements, this hospital was declared as the Focal Point for National Quality Assurance Program by the country's Ministry of Health. Since then senior and middle-level managers of other hospitals in charge of quality improvement have been trained at Castle Street Hospital for Women. As such many other hospitals in Sri Lanka followed the pathway of Castle Street Hospital for Women to improve quality and safety in their hospitals (JICA 2008).

Before   **After**

Fig. 8.1   Storage of medical records

From 2007 with the assistance from Japan International Cooperation Agency and Japan Medical School, Karandagoda and his Japanese colleagues began standardizing 5S-KAIZEN methodology for the counterpart hospitals in India, Bangladesh, Thailand, and African countries (Fig. 8.1).

3   Story Two: Tanzania

Tanzania has suffered from a serious shortage of human resources in the field of healthcare (The World Health Report 2006). As of 2010, public medical institutions had only about 40 % of the personnel they needed, and more than 20,000 specialist posts remained unfilled. A lack of resources overall, as represented by that of personnel, is said to be the source of the nonexistent improvement in quality of services provided by healthcare facilities.

When "Clean Hospital Program" was launched in 2007, Dr. E.R. Samky, director general of Mbeya Referral Hospital, was the first person that raised his hand to participate in the program. After visiting Castle Street Hospital for Women in Sri Lanka, he was convinced with the expected outcomes that 5S would bring about to his hospital. Consequently he initiated quality improvement teams and operation improvement teams in every ward of his hospital. A number of staff members refused cooperation, questioning why they, specialists, had to take part in a house cleaning campaign. Nevertheless the quality improvement team carried on the activities slowly and patiently. The results became obvious: they were able to reduce the amount of dead stock of medical supplies by 37 %. This led to increased level of perceived solidarity among all the employees. In 2010, Mbeya Referral Hospital was ranked the number one in the national hospital audit.

Fig. 8.2 Training session in fishbone chart (Photo: provided by Hisahiro Ishijima)

Currently 5S-KAIZEN movement has diffused throughout the country. As of 2011, 46 hospitals in Tanzania were using the 5S-KAIZEN methodology. The Tanzanian government has incorporated the 5S-KAIZEN methodology into its public health policy and management (Ishijima 2012) (Fig. 8.2).

4   Story Three: Congo

GDP per capita of Democratic Republic of the Congo was US$ 348 (International Monetary Fund Data and Statistics 2009). As such this country remains one of the poorest of the world with very limited access to the fee-for-services even in its capital city Kinshasa. After observing the 5S practices of African neighbors, the government appointed clinic Ngaliema Hospital, a national hospital, as a pilot institution to evaluate the effectiveness of that method. Then, Dr. Chamara, president of the hospital, organized work improvement teams and quality improvement team in his hospital.

A humble example of improvement is medical record administration. The medical records were in a state of disorderliness and mess. But after the 5S efforts they were kept by years and patient names. Then newly appointed medical record manager began visualizing basic health statistics by using such database. Other outcomes included the decreased waiting time for outpatients, increased quality of services, and increased revenue, which enables the president to pay the accrued

Fig. 8.3 Usage of medical statistics (Photos: provided by Noriaki Ikeda)

wage and purchase a limited number of medical equipment (Ikeda 2012). In 2011 the Ministry of Health incorporated 5S-KAIZEN-TQM as a pillar of national health management policy. Sixteen hospitals in the state of Kinshasa are in the process to installing 5S-KAIZEN programs into their institutions (Fig. 8.3).

5   KAIZEN and Innovation

Imai (1997) argues that "the word [KAIZEN] implies improvement that involves everyone—both managers and workers—and entails relatively little expense. The KAIZEN philosophy assumes that our way of life—be it our working life, our social life, our home life—should focus on constant-improvement efforts." KAIZEN or continuous improvement is apparently different from innovation. KAIZEN is geared towards incremental change, alteration, transition, and transformation rather than disruptive changes or innovation.

The changes, even though they are disruptive or incremental and large scale or small scale, tend to emerge in three evolutionary stages. Those include such phases as products/goods, processes, and services. With servitization manifesting itself in various industries, decision makers begin allocating and investing resources to realize changes in service offerings. These days, as a result, more attention is being paid to changes in service sectors which include health and medical services (Fig. 8.4).

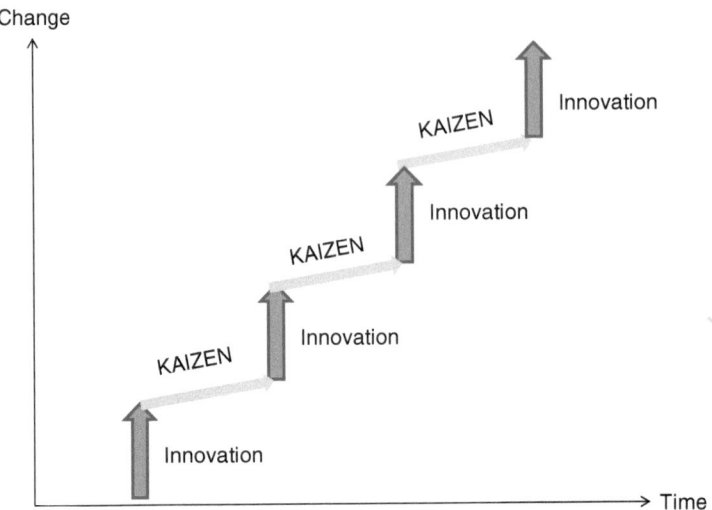

Fig. 8.4 KAIZEN and innovation. *Source*: presentation prepared by Karandagoda. Sri Lanka 2011

The relationship between innovation and KAIZEN is illustrated in the above figure. Innovation tends to bring about radical and disruptive changes in a short period of time, whereas KAIZEN is inclined to generate incremental and gradual changes in relatively longer period of time. Technological innovation in particular can be emerged by groups of elites with abundant operant resources, whereas KAIZEN can be performed by everyone. However, innovation and KAIZEN are not mutually exclusive phenomena, but they are dependent when they are observed from a long-term phenomenological point of view.

6   Translational Aspects of 5S-KAIZEN

Today the term "translational research" is frequently used in medical and healthcare. The process of translating basic scientific discoveries into clinical applications, and ultimately into public health improvements, has emerged as a salient but complex objective in biomedical research. Here the process can be described as a "translation continuum," since various resources, actions, and processes are involved in this progression of knowledge, which advances discoveries from the "bench to the bedside" (Wolf 1974). Scientific discoveries are translated into practical applications to improve human health. Such discoveries typically emerge at "the bench" with basic research where scientists study disease and then progress to the clinical level or the patients' "bedside." The bench-to-bedside approach to translational research is a two-way co-creation in the translation continuum. In bench-to-bedside, scientists provide practitioners with new tools for practical use in patients and for assessment of their impact. In bedside-to-bench, on the other hand, clinical practitioners make novel observations about the disease that often stimulate further scientific research and exploration.

Based on the aforesaid notion, however, this chapter uses "translational approach" to mean a new kind of approach with the capacity to tackle challenges by sharing disciplines and by developing *new continuum* that combines different sets of agents and relations in an attempt to emerge a new system. To put simply, in translational approach, new continuum has emerged as a human activity system (S):

$$S = (a,r)$$

where a = agent, r = relation

Based on the above, there are at least four distinct elements of translational approach in 5S-KAIZEN:

(1) Bridging Effects on Products/Goods, Processes, and Services

First, originally developed and utilized widely in the Japanese manufacturing industry, 5S-KAIZEN as a quality control and improvement methodology has been transferred to and adapted by the health services sectors. Nowadays more than 65 % of the 9,000 of Japanese hospitals are using methodologies to improve quality of health services, work environment, risk management, and safety. Thus as a methodology per se, 5S-KAIZEN has covered and come across from manufacturing sector producing products/goods to service sector including health services. 5S-KAIZEN is translational in that it has been effectively transferred, adapted, and utilized as a management tool by various agents in different industries and sectors.

5S-KAIZEN has translational effects to bridge three different spheres involving products/goods, processes, and services. 5S-KAIZEN is able to improve, integrate, and translate the continuum that involves such artifacts as products-goods, processes, and services in various healthcare settings. Those include clinical bedside settings, health services administration of clinics and hospitals, institutions related to health policy, and management.

(2) Adaptability to Multiple Cultures

Second, 5S-KAIZEN, as a tool to improve the levels of combination of health services, work environment, risk management, and safety, has been translated from its home country to African countries. It has created new relations that did not exist before. It is proliferating even to the hinterland of Africa covering 46 countries, impacting on directly or indirectly the human life of 420 million people, and accounting almost half of the population of 820 million. 5S-KAIZEN is translational in that it has been implemented by various agents in different countries with multiple cultures that include: Eritrea, Kenya, Madagascar, Malawi, Nigeria, Senegal, Tanzania, Benin, Burkina Faso, Burundi, the People's Republic of the Congo, Mali, Morocco, and Niger (Hasegawa 2012).

Japanese	English	French	Spanish	Yemen	Egyptian	Russian
整理(せいり)	Sort	Séparer	Clasificar	التصنيف	•افرز	Упорядочение
整頓(せいとん)	Set	Situer	Organizar	الترتيب	•مرتب	Организация
清掃(せいそう)	Shine	Salubrité	Limpiar	التلميع	•المع	Опрятность
清潔(せいけつ)	Standardize	Standardiser	Estandarizar	المعايره	•اجعلها مثالا	Чистота
躾(しつけ)	Sustain	Se Discipliner	Mantener	الاستمرارية	•محافظ	Поддержание

Shinhala	Swahili	Malaysian	Malagasy	Burundian	Senegal
සකසමු (කංවීඩ ඉවි)	Sasambua	Sisih	Sivanina	Kuvangura	Supprimer
සකසතා සුවිවිදිව	Seti	Susun	Sokajiana	Kutondekanya	Systématiser
සබිතා (සුවිවිදිකම)	Safisha	Sapu	Sasana	Kubungabunga	Scintiller
සබිතාමුදු (සවිවිකම)	Sanifisha	Seragam	Soratana	Kumenyera	
ඉවිගුවි (ඔවිවඟත)	Shikilia	Sentiasa Amal	Saintsainina	Kwamizako	Suivre

Fig. 8.5 Translations of 5S. *Source*: presentation material made by Hasegawa (2011)

As shown below, for instance, 5S have been translated into African local languages. Translation literally refers to the action of turning from one language to another. At the same time translation enables one to deeply comprehend the meaning embedded in different languages. Indeed translation is the process to enhance the diffusion of the methodology across and within diversified cultural boundaries. As such semantic translationality realized by translation into multiple local languages has enabled to further transfer, remove, and convey from one person, place, or condition to another with necessary alteration, change, and adaptation (Fig. 8.5).

(3) Common Tool

Third, 5S-KAIZEN provides a common tool of change for different professional agents including physician, dentist, nurse, public health nurse, midwife, nutritionist, pharmacist, medical technologist, radiological technologist, clinical engineer, physical therapist, occupational therapist, and office clerk, to name just a few. Usually the medical and health service professionals are expected to develop and utilize relatively narrow scope of clinical disciplines in their specialized areas. This tendency unfortunately results in the lack of common language bridging sets of different expertise which leads to difficulties in realizing changes in cross-functional settings.

In such circumstances, however, 5S-KAIZEN has come to function to build and maintain cross-expertise relations within a variety of professional agents. A number of people with different clinical backgrounds are now able to find, define, and improve problematic situations using this methodology (Fig. 8.6).

(4) Open and Free

Fourth, changing in today's environment requires us to be open. Openness can reduce the costs of access to knowledge and information, costs of transaction,

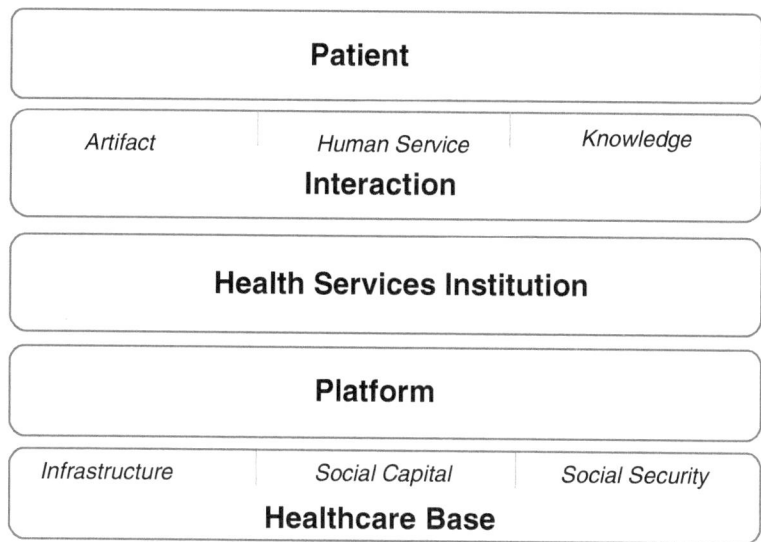

Fig. 8.6 Hierarchical structure of health services

and costs of change. This helps to share the risks and rewards of changes and accelerate the time required to deliver the changed outcomes to communities. The contents relating with the method and methodology of 5S-KAIZEN are open and freely used and shared by the concerned agents.

Generally the intellectual properties of methods or methodologies concerning with the newly discovered or developed in bench-to-bed translational settings tend to be protected and claimed exclusively. Such discoverer and developer have a distinct intention to secure their rights in order to safeguard their invested money, future opportunities to acquire profits, or at least the usage of their intellectual properties. Except for the simplistic copyrights of documents, however, the intellectual property of the method and methodology concerning 5S-KAIZEN has not been exclusively claimed by any individuals or institutions for the purpose of acquiring or gaining monetary profits. As such various agents, including individuals and institutions, are able to access, utilize the methodology pertaining to 5S-KAIZEN openly and freely, and create new relations involving new agents as well.

7   Embedded Systemicity

Given the translational aspects of 5S-KAIZEN as discussed above, 5S-KAIZEN methodology and its underlying philosophy could bear universality which goes beyond the boundaries restricted by national, cultural, and industrial boarders.

I arguably assert that the *translationality* derives from the embedded *systemicity* of 5S-KAIZEN, of which system property constitutes three dimensions, i.e., (1) holistic and hierarchical, (2) communication and control, and (3) evolution (Kijima and Jackson 2007):

(1) Holistic Changes

5S-KAIZEN is able to bring about holistic changes to each of the layers of the hierarchical structure of health services as illustrated below. Many of the institutions, including health and medical teams, its subgroups, clinics, hospitals, medical centers, home care deliverers, and community day care centers, to name just a few, interact with patients in order to co-create health services through using such artifacts as medicine and medical device, human care, and knowledge. As is explained and evidenced in the following chapters, numerous healthcare institutions have found 5S-KAIZEN efficient and effective for increasing not only levels of quality of care, safety, and patients' satisfaction but also employees' job satisfaction. These cases are associated with such layers of patients, health services institutions, and interaction.

These institutions on the other hand depend upon platform layer. Despite the differences in health policy in countries, platform functions as a frame of a healthcare system's deliverables, defines how healthcare institutions are operated, and determines what kinds of resources should be allocated and utilized. Therefore, platform, when it is appropriately aligned, enables community of practices to transcend disciplinary boundaries towards developing new perspectives concerning knowledge, human services, and development and application of products/goods. Platform is largely restricted by healthcare base which constitutes of infrastructure, social capital, and social security systems (Fig. 8.7).

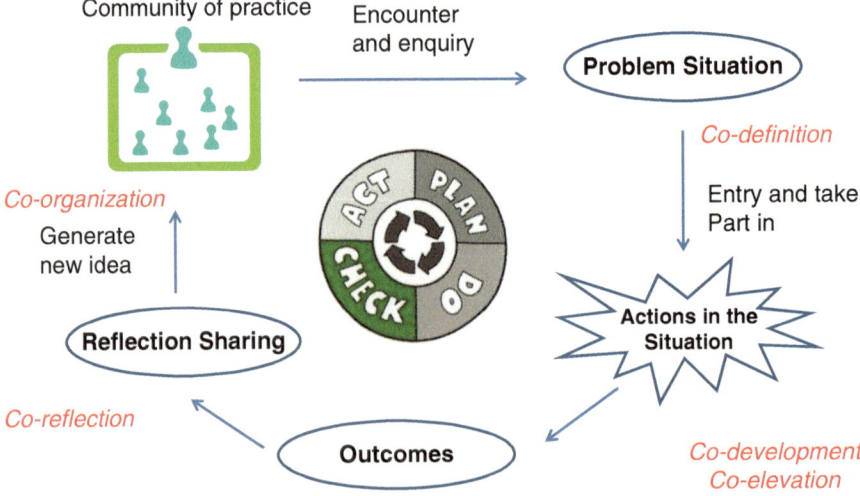

Fig. 8.7 Value co-creation in action research-based learning

Observing that the policy makers have officially introduced 5S-KAIZEN as the national health policy and that they have recognized its effectiveness in such countries as Tanzania and Democratic Republic of the Congo, this methodology has impact, if limited, on platform and healthcare base. Consequently 5S-KAIZEN methodology has realized holistic changes directly or indirectly to the combination of each layer above.

(2) Communication and Control

After introducing 5S-KAIZEN a number of institutions have reported the increased degree of internal communication, control, and participation. In an organizational context learning occurs when a person is able to control their learning experience and the persons sharing contents and contexts are able to communicate their learning experiences with each other.

The concept of "community of practices" was first proposed by Lave and Wenger (Lave and Wenger 1991). It is through the process of sharing knowledge and experiences with the group that the members learn from each other and have an opportunity to develop themselves personally and professionally (Fig. 8.8).

Action research is a reflective process of progressive problem solving led by individuals working with others in teams or as part of a "community of practice" to improve the way they address issues and solve problems. The action research program that was established at Lancaster and yielded these early successes has since been used in hundreds of projects (Jackson 2003). An action

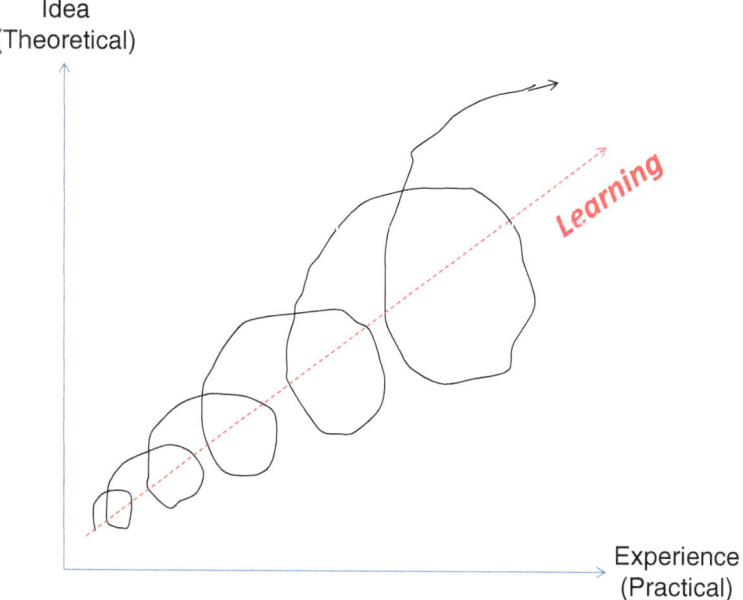

Fig. 8.8 Evolutionary learning based on 5S-KAIZEN

research program, one of the representative approaches in soft systems methodologies, enables the co-creative circulation of interactions between cognition
and experience (Kijima 2007).

Figure 8.4 illustrates value co-creation flow in 5S-KAIZEN action research-
based learning. In using 5S-KAIZEN methodology practitioners are expected
to co-define problem situation after encountering with the situation. Then they
take part in actions in the problem situations by codeveloping new relations
and co-elevating solutions focusing on entity and agent. Following those steps
each of the participants and the communities are able to share reflections (co-
reflection) by evaluating the outcomes of their intervention. When this collective reflection is satisfactory enough, each member of the community of
practice is able to generate and obtain new ideas to further intervene the situations.
By so doing communities of practice are to co-organize such operant resources
as human resources, materials, money, knowledge and information, space, and
time (co-organization).

Consequently 5S-KAIZEN methodology is able to enhance not only communication and control but also personal and transpersonal learning through
recursive cycle consisting of co-definition of the problem situation, codevelopment and co-creation of solutions, and co-reflection and co-organization of
operant resources.

(3) Evolution

Let us hereby assume the element of time to the above discussions. The process
of learning in nature runs parallel with time. Learning in a dynamic context, that
is, learning in the consideration of time series, can be expressed as a recursive
movement between experience (practice) and idea (theory) through the processes of co-experience involving co-definition, codevelopment (relation-
based), co-elevation (agent-based), co-reflection, and co-organization as
illustrated below (Fig. 8.9).

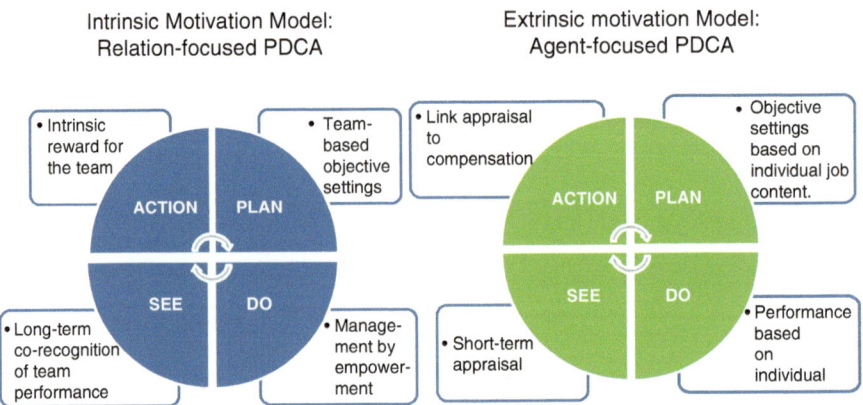

Fig. 8.9 Relation-focused and agent-focused PDCA cycles

In situations where "the notion of 'problem' and a 'solution' are inappropriate, what makes more sense is a process of learning which is never-ending" (Chambers 1997). Therefore what is essential here is an "inquiry" in the part of practitioners when they encounter problem situations rather than independent and contained problems "out there." Never-ending but effective inquiry to grasp and describe the problem situations ensures the process of co-experience.

As is shown in the following chapters, successful practices of 5S-KAIZEN incorporate the learning processes coherently, where each of the participants acts as an actor to enhance the recursive and evolutionary cycle between experience and idea.

8   Behind the PDCA Cycle

The *systemic* changes realized by the movement have gained enthusiastic support from the minister down to the staff members at hospitals and clinics in countryside in Africa. African people have even produced "5S-KAIZEN dance" and "5S-KAIZEN song" to encourage themselves to get this movement forward. Here a question is raised. Why are African people so enthusiastic about "5S-KAIZEN"? It seems that what drives plan-do-check-act cycle forward really matters.

(1) Participation and Inclusion

The management ethos has been excluded in health services partly because of the poverty and inadequate education. Even stealing medicine or equipment is not rare in African countries. But this approach has the effect of changing passive mind-set to proactive one. In particular, everyone can do the 3Ss or sort, set, and shine easily and assure the outcomes. In this sense the participation in 5S-KAIZEN sets the easy but steady start line to introduce the discipline of self-management.

Social exclusion is a multidimensional process of progressive social rupture, detaching groups and individuals from social relations and institutions and preventing them from full participation in the normal, normatively prescribed activities of the society in which they live (Silver 2007). On the other hand social inclusion, often being an agenda of debate, could be perceived as a set of actions to change the circumstances and habits that have led to social exclusion. Usually practiced within a certain institutional boundary, 5S-KAIZEN facilitates the processes of such actions as participation and inclusion within the social context relating the institutions.

(2) Intrinsic Reward

Most of the management methodologies and methods if any currently practiced in African countries have come from the West which historically had exploited the continent for decades. In those practices they try to give reward for individuals and organizations based on meritocracy, not to say carrot and stick. Those practices by and large are based on goal-seeking paradigm; they plan in advance the acted-out results of individuals and organizations that should be achieved.

When those are achieved, the incentives, including monetary or subsidies, are to be awarded to those who achieve. For those who are sick and tired of such ways, emotional welling up from the inner sense of fulfillment and a sense of accomplishment based on relation-oriented paradigm means intrinsic reward. 5S-KAIZEN brings about the intrinsic reward.

(3) Solidarity

"Ba," a Japanese word, means a place or field associated with human activities. According to Nonaka and Konno (1998), "Ba" can be thought of as a shared space for emerging relationships. This space can be physical, virtual, mental (e.g., shared experiences, ideas, ideals), or any combination of them. "Ba" provides a platform for emerging relations and learning among the concerned.

What supports and integrates such "Ba" as workplace and community is the collective sense of solidarity which refers to the ties or integration in a micro society that binds persons to one another. When such sense of solidarity is enhanced through planning, doing, checking, and doing actions based on 5S-KAIZEN, people can assure and even enrich solidarity at workplaces.

The PDCA cycle, when it is appropriately introduced and operated, has helped the participants in 5S-KAIZEN realize participation, intrinsic reward, and solidarity in human activities related with health services.

9   Conclusion

This essay was an attempt to describe some aspects of methodology of operations management, i.e., 5S-KAIZEN practices in Asian-African transnational and translational community of practice in value co-creation in health services. Although the African practices of 5S-KAIZEN seem humble and somewhat far away from sophistication, they imply some crucial perspectives to systems science and service systems management. Systems science has traditionally emphasized the importance of crossover and mash-up of disciplines by frequently using such words as multidiscipline, inter-discipline, and trans-discipline. But as far as human activities are concerned there has been virtually no concrete methodology if any to realize multidiscipline, inter-discipline, and trans-discipline practices.

This chapter provided some implications when it comes to the attempt to realize such practices. In order for us to realize multidiscipline, inter-discipline, and trans-discipline in certain problematic issues, we should be able to put the issue in the translational continuum of idea and practice which involves *episteme* (concept and logic), *techne* (theory and model), and *prophenis* (practice). Seen from a dynamic viewpoint, the translational continuum runs parallel with recursive learning cycle which in turn facilitates crossover and mash-up of disciplines.

People at the service forefront especially in health services may benefit from easily and steadily practicing 5S-KAIZEN. Even though such word as value co-creation is somewhat used as a buzzword these days, we may be able to utilize 5S-KAIZEN as a soft systems methodology for value co-creation in a concrete manner. This may provide us with new perspective to service systems management of such human activity system as health services.

References

Chambers R (1997) Whose reality counts? Putting the first last. Intermediate Technology, London

Handa Y (2012) Changing hospitals in Africa. Hospital, Igakushoin 71(2):598–599

Hasegawa T (2011) Changing hospitals in Africa. Hospital, Igakushoin, (Japanese), Tokyo

Hasegawa T (2012) Changing hospital: the present Africa. Hospital, Igakushoin 17(1):942–943

Ikeda N (2012) Changing hospital in Africa. Hospital 71(10):342–343

Imai M (1997) Gemba Kaizen: A commonsense, low-cost approach to management. McGraw-Hill, New York

International Monetary Fund Data and Statistics (2009) Democratic republic of the Congo: statistical appendix. Archived at: http://www.imf.org/external/pubs/ft/scr/2010/cr1011.pdf

Ishijima H (2012) Benchmark hospital in Africa. Hospital 71(4):430–431

Jackson M (2003) Systems thinking: creative holism for managers. Wiley, West Sussex

JICA Bulletin (2008) Castle street hospital sets an example to the Asia –African Regions. http://www.jica.go.jp/srilanka/english/office/others/pdf/brochure108.pdf

Kijima K (2007) Holistic creative management. Maruzen, p 96 (Japanese), Tokyo

Kijima K, Jackson M (2007) Systems thinking. Holistic creative management. Maruzen, England

Lave J, Wenger E (1991) Situated learning: legitimate peripheral participation. Cambridge University Press, Cambridge

Matsushita H (2013) Chapter I-3: 5S-KAIZEN-TQM as a methodology of value co-creation: implications for health services management. In: Japan International Cooperation Agency, Change Management for Hospitals (project document), pp 4–10, Tokyo

Nonaka I, Konno N (1998) The concept of 'Ba': building foundation for Knowledge Creation. Calif Manag Rev 40(3):40–54

Silver (2007) Social exclusion: comparative analysis of Europe and Middle East Youth. Middle East Youth Initiative working paper. p 15

The World Health Report (2006) http://www.who.int/whr/2006/en/

Wolf S (1974) The real gap between bench and bedside. New Engl J Med 290:802–803

Chapter 9
Creating Information-Based Customer Value with Service Systems in Retailing

Timo Rintamäki and Lasse Mitronen

Abstract With the advent of mobile technology, addressing the information needs of customers across channels has become a key source for value creation. Also, this information-based value creation has implications for how retailers design and manage their customer value propositions for competitive advantage. As our data from the USA, Japan, and Finland show, shoppers already use multiple channels for their prepurchase, purchase, and post-purchase activities. Understanding the roles of different channels in the individual stages of the customer experience provides valuable input for service system development. Those retailers who have mastered the planning of service systems and consider their implications for information-based value creation can avoid being stuck as an endpoint of logistics.

Keywords Customer value propositions • Information value • Multichannel retailing • Service systems

1 Introduction

Across industries, new ways to create competitive advantage are often found in the emerging innovative uses of information. And this time, the basis for the competitive advantage is not a scarce resource. Quite the contrary, in most modern organizations, the volume of data is expanding by 35–50 % every year. This is represented by the need to process 60 terabytes of information annually—a thousand times more than a decade ago (Beath et al. 2012). What is scarce is competence to harvest the benefits of information, especially in a way that benefits the customer. Thaler and Tucker (2013) have coined the term "choice engine" to describe technologies that enable vast quantities of data to become meaningful and timely information for customers and for citizens in general. These "choice engines" are made possible by

T. Rintamäki (✉)
University of Tampere School of Management, Tampere, Finland
e-mail: Timo.Rintamaki@staff.uta.fi

L. Mitronen
Aalto University School of Business, Helsinki, Finland

© Springer Japan 2015
K. Kijima (ed.), *Service Systems Science*, Translational Systems Sciences 2,
DOI 10.1007/978-4-431-54267-4_9

the smart disclosure of government- and company-maintained information and/or usage data in machine-readable form. As Thaler and Tucker show, both the public and the private sector have vast possibilities to contribute in this respect. On the public side, initiatives such as data.gov in the USA and the UK's data.gov.uk release datasets and invite private-sector organizations to develop applications and tools, with the results including better services for citizens and also opportunities for existing and new businesses. Private-sector companies are also strongly in the game of smart disclosure. Tesco, for instance, has plans to offer its loyalty-card users planning and goal-setting functions based on shopper histories—showing new thinking on utilization of usage data. The source of the data notwithstanding, we believe that these examples witness times of a new logic for value creation: information-based value creation.

However, there are often wide gaps between raw data, the technical solution, and a happy customer. Bridging the scattered bits of decision data and the actual contextual decisions requires not only accessible information and technology but also access to customers' contexts and meaningful interactions between customers and the organization. In other words, a service system (see, e.g., Mele and Polese 2011) is needed if smart disclosure and "choice engines" are to be possible. While many organizations struggle with this challenge, some innovative incumbents, along with numerous start-ups, focus on designing and managing service systems that use this information to enhance value creation for their customers. Perhaps surprisingly, we see retailing as one of the areas wherein the change has been—and can continue to be—truly translational: it is an area that is strongly related to the everyday needs and wants of consumers, offering plenty of potential for helping people in day-to-day life, and it requires a multidisciplinary approach, for finding new solutions based on service logic and the value of timely information.

To this end, we set out to explore here how information-based customer value can be created in retailing and what kinds of challenges and possibilities it offers for service systems' design. Theoretical background is organized around the discussion of value creation in a retail context and the role of service system design in its facilitation. In order to illustrate the key concepts and ideas, we use shopper data based on surveys of smartphone users from the USA, Japan, and Finland, putting the focus on how customers use offline, online, and mobile channels in their prepurchase, purchase, and post-purchase activities. The mobile channel is given special emphasis because it has a potential role in bridging the offline and online channels. Differences among the three countries are presented, and the results are used as identifying criteria for service system design across service channel boundaries.

2 Retail Transformation Through Service

Theoretical background for this paper is presented in two subsections. In the first of these, service science and service-dominant logic are suggested as a theoretical perspective for understanding the recent developments in retailing that manifest themselves through customer-centric development of information systems and channel integration. The second subsection addresses the value creation in the multichannel environment, where the information is utilized across channels for the benefit of the customer. The resulting cross-channel retail environment provides a context and tools for information-based value creation in retailing. Value is a key phenomenon and construct, and, hence, its creation is approached from customers' and not merely the retailer's perspective. Argument is made also that there is a role for information in value creation.

2.1 Service Thinking as a Driver for Repositioning the Role of Information Systems in Retailing

As in many other areas of business, information systems have had a key role in retail innovations as significant changes have taken place in the operating environment. However, the role of information systems has until recently been based mainly on inside-out logic, focused on how production-oriented resources and activities can be more efficiently and effectively managed through information and technology. In other words, information systems have enabled more efficient backstage processes for retailers and their value chain. For instance, the developments that are characteristic of the entire field have comprised new ICT for international procurement for utilization of both electronic and real-time information transfer between organizations (EDI) and also for the utilization of product-specific information (POS/EPOS and e-commerce systems). Other noteworthy information-exchange-based innovations have included Efficient Consumer Response (ECR), the implementation of product group management (category management), the shifting and merging of roles in retail and wholesale trade, mergers of trade companies, and the increased importance of customer relations (customer relationship management, or CRM) as part of integrated marketing. As a common feature, the role of the customer in the information system has remained somewhat passive.

As a more recent development, outside-in thinking has gained stronger impetus among the innovators of retailing, with some information systems having been opened up for retail customers too—i.e., shoppers. Instead of focusing on the strengths and the prevalent modus operandi of the focal organization, the inside-out strategist looks at the market and asks questions such as "how can we deliver new value to our customers?" and "what new capabilities do we need?" (Day and Moorman 2010, 5). Indications of these new ways to open up and further develop information systems to create new value for shoppers can already be seen in many retailers' operations: customers can see real-time inventory status, access their loyalty-card/consumption data, and order and pay for products online. However, the real change can be seen as some leading incumbents and agile newcomers build their key competencies on outside-in thinking: instead of being the endpoints of logistics, they develop new information-based service models. This requires an in-depth understanding of how customers select and use their products, alongside development of ways to support customers in these processes. For instance, consider how Amazon.com and the Japanese Rakuten have systematically developed tools for customers to find, compare, select, experience, and review products. Then think about how many information systems have been integrated to achieve this and what a different mindset and set of competencies it has required.

The shift from inside-out to outside-in thinking and the emphasis on information and new capabilities can be identified and argued for from the perspective of service-dominant logic (Vargo and Lusch 2004, 2008) and service science (management, methods, engineering, and design) (Maglio and Spohrer 2008). For these streams of research, service is about applying information and skills as the key competencies for the benefit of another party (Vargo and Lusch 2004, 2008) and about co-creation of value among multiple entities (Spohrer and Maglio 2010; Demirkan et al. 2011)—in other words, to do with how customers, other people (employees and other stakeholder groups), information, and technology can be managed as a service system for value co-creation (Mele and Polese 2011; see also Maglio et al. 2009). When the information systems are designed and managed so as to create value for the customer, they have potential to form the foundation for a modern service system.

Perhaps the most important and concrete change in how shoppers shop and how retailers do business has been the introduction of multichannel and cross-channel (and omni-channel) retailing (Deloitte 2012; Rigby 2011; Zhang et al. 2010). Resmini and Rosati (2011, p. 54) illustrate this thus: "Everyday shopping does not concern itself with the convenience store or supermarket only, but configures a process that may start on traditional media, include the Web, proceed to another shop to finalize the purchase, and finally return to the Web for assistance, updates, customization and networking with other people or devices." Compared to traditional multichannel retailing, cross-channel solutions are aimed at synergy and coherence among the channels. Moreover, channel integration for seamless shopping experiences calls for resources that utilize the dynamics between channels, whether they are channels for sales, communication channels, or tools for decision-making. Chatterjee (2010, 10) explains the difference between multi- and cross-channel as involving two separate strategic options: operate multiple channels as independent entities (in a multichannel strategy—i.e., order and pick up in-store, order online or by telephone, and get the product delivered) or integrate multiple channels, allowing cross-channel movements of products, money, and information (in a cross-channel strategy—order online/pick up in the store, order in-store, and get the product home-delivered). When the multichannel solutions are designed and integrated from the customer's perspective, cross-channel service is created.

Cross-channel retail service can be defined as a service system, because it consists of at least a pair of entities (e.g., a provider and customer), their interactions, and sharing of access to configure resources for mutual value creation (Spohrer and Maglio 2010; Demirkan et al. 2011). Of the four dimensions of a service system as proposed by Mele and Polese (2011), the role of technology and that of information are emphasized in the cross-channel retail context. Not only the quantity of data poses a challenge; its nature does too: data are often unstructured and stored in various formats, so they are not easily interpreted or retrieved. Beath et al. (2012) give examples of the explosion of both structured and unstructured data. The body of structured data grows as organizations introduce and develop systems such as enterprise resource planning systems, management systems, CRM systems, and technologies such as RFID. Unstructured data, on the other hand, are all about scattered materials, e-mail messages, videos, images, etc. Beath et al. (2012, p. 18) note that "at most organizations, generating business value from increased amounts of data is still an aspiration." However, this type of data is often cocreated with customers (Novani and Kijima 2012). As a context, social media has seen the development of many algorithms designed to make sense of unstructured data. For instance, Walmart's Shopycat application experimented with an algorithm on Facebook that analyzed social-media discussions and, from the results, matched gift suggestions to the friends of Walmart shoppers.

2.2 Customer Value Propositions in the Age of Information

The value proposition is a key concept for understanding service systems and an essential element of customer-oriented retail strategy for competitive advantage (Payne and Frow 2014). In the context of service systems, a value proposition is a value co-creation mechanism that communicates a mutually agreeable plan to collaborate and cocreate value (Spohrer and Kwan 2009). The retailer has an important role as a resource integrator who links in consumer customers, just as much as it does the collaborating companies and other stakeholders, around a common value proposition (Lusch et al. 2007). The value proposition also provides motivation for all parties to reconfigure the actual resources or access rights to resources. In addition, it acts as the "glue" between the systems, making it possible to refine new value by connecting existing databases, as in the case of Amazon (e.g., synthesizing product data, content rating data, and customer review data with customers' preference and purchase-history data).

When brought into the retail strategy domain, value propositions (specifically, customer value propositions) describe a market positioning for competitive advantage (Anderson et al. 2006). A sound customer value proposition should increase the benefits and/or decrease the sacrifices that customers perceive, build on those competencies and resources that can be better utilized than competitors', be unique and hence recognizably different from the competition's, and result in competitive advantage (Rintamäki et al. 2007; see also Webster 1994). Hence a competitive customer value proposition steers the use of resources and competencies in addition to marking a market positioning in the minds of the customers.

In the context of retailing, Rintamäki et al. (2007) identify customer value propositions along four dimensions of customer value: economic, functional, emotional, and symbolic value. Economic value is judged in monetary evaluation of products and services. Hence, the focus is on price. Functional value is expressed as savings on time, physical effort, and the cognitive cost of decision-making. The goal is also to make sure that the right customers end up with the right products and services. Accordingly, it also encompasses traditional perspectives on product quality and product selection choices. Functional value thereby moves the focus from price to solutions. Emotional value results from the positive feelings and emotions that products and services stimulate. Accordingly, the focus is on the evaluation of the customer experience: how to build preference and add stimulation and/or enjoyment of shopping. Symbolic value moves the focus to the meanings that products and services represent. For instance, buying ethically produced goods or patronizing an upscale retailer might yield symbolic value.

We believe that cross-channel retailing, along with the corresponding changes in consumer behaviors, calls for a new logic for delivering the proposed value. This is especially due to the possibilities that online and mobile technologies offer in various stages of the shopping experience. As Peterson et al. (2010) note, shoppers' decision processes related to what to buy and where to buy have gone digital (Peterson et al. 2010). The digital tools may be provided by retailers themselves in the form of Web pages and mobile "apps" or by third parties. In any case, the shift entails more and more decisions on what to buy and where to buy being made outside the brick-and-mortar stores. Besides the prepurchase stage, these digital tools may be used in the actual purchase stage and in the post-purchase stage. Hence the new logic for value must lie in the ability to serve customers with information across channels.

The literature on information as a source of value is rather scattered. Proposing a definition for the information value of virtual communities, Archer-Brown et al. (2013) identified Hirshleifer (1973) as one of their key sources. According to them, the value of information lies in five assumptions: certainty (ability to resolve uncertainty), diffusion (ease of distribution), applicability (how the receiver can apply it), content (the nature of the information), and decision-relevance (how the receiver can use it in decision-making). These basic assumptions also hold in the retail context, where decisions about what to buy and where to do so are made in conditions that are often rendered surprisingly complex by ever-increasing choice. Resmini and Rosati (2011) identify five heuristics for a pervasive information architecture, which also can be used here to characterize information-based value creation: (1) place-making, (2) consistency, (3) resilience, (4) reduction, and (5) correlation. Place-making is about reducing disorientation, making sense of place, increasing legibility, and supporting way-finding across channels. Consistency refers to both internal and external demands. Internally, consistency involves suiting the purposes, contexts, and people, whereas external consistency entails maintaining uniform logic across boundaries of media, environments, and time. For our purposes, this is a key characteristic for making sure the information is both available and applicable across channels and in different stages of the customer experience. Resilience is attained when the system is able to shape and adapt itself in line with specific user needs and seeking strategies. In the context of retailing, resilience might be seen in how the retailer can learn from customers' purchase history and preferences across channels. Although customization to customer-specific preferences can be "automatic," in many cases, digital retail concepts also provide tools allowing customers to make conscious decisions on how to customize the service. Reduction is facilitated through supporting the user in managing large information sets and hence reducing potential frustration and stress of choosing from among vast quantities of services and goods representing different sources of information. Finally, correlation refers to the ability of the system to suggest relevant connections and thereby help users be aware of their needs or meet their latent needs via the pieces of information, services, and products. For instance, Zappos has experimented with several ways to curate styles and trends across product categories with applications such as Glance by Zappos.

To unleash the value of information across channels as described by Resmini and Rosati (2011), many retailers have turned to mobile technologies for their ability to bridge channel behaviors. Emphasizing the role of mobile technologies and social media, Larivière et al. (2013, pp. 277–278) provide the following definition of value, with what they call value fusion: "Value that can be achieved for the entire network of consumers and firms simultaneously, just by being on the mobile network. Value Fusion results from producers and consumers (i) individually or collectively, (ii) actively and passively, (iii) concurrently, (iv) interactively or in aggregation contributing to a mobile network (v) in real time and (vi) just in time." Lariviére and colleagues add some important elements to our understanding of information value: it is often continuously updated, jointly created, and maintained by retailers, customers, and/or third parties, and it is always accessible. Hence, the term "value fusion" inherently suggests the importance of the unstructured data and the utilization of said data in mobile value creation (Beath et al. 2012).

In the context of vacation travel, Cho and Jang (2008) conceptualize and measure information value along five dimensions: utilitarian, risk-avoidance, hedonic, sensation-seeking, and social. Although our context is different, they provide an important conclusion for conceptualizing information value: instead of trying to isolate it as a new or distinct dimension of value, one may be better off investigating how existing dimensions of value are related to the idea of creating information-based value. In other words, all perceptions are based on sensory information, but information-based value creation can be defined as a systematic way to provide customer information that supports customers' value-creating processes.

For the purposes of this paper, we offer the following definition for information-based value creation: *Information-based value creation systematically refines and combines contextual data on acceptable prices, customized solutions, preferred experiences, and/or personalized meanings to support customers' prepurchase, purchase, and/or use processes.* We believe information-based value creation marks a new type of frontier for customer-centric value creation, where the retailer is more than the endpoint of logistics and where the active role of the customer as a coproducer or cocreator of value is acknowledged.

3 Exploring Information-Based Value Creation in Retailing

To gain understanding of how customers behave in the prepurchase, purchase, and post-purchase stages of shopping and to see how they use their mobile devices to bridge channels, we conducted an online survey in summer 2012 in Japan, the USA, and Finland. The survey was completed as part of a two-year research project wherein 200 mini-cases were systematically recorded that illustrated multi- and cross-channel behaviors in retailing. These cases were also used, in addition to literature on cross-channel developments, in questionnaire development. We further illustrate information-based value creation in the three main stages with selected examples based on some of the mini-cases.

3.1 Survey Data

A professional research agency was used in sampling and data collection. For all three countries, the goal in the sampling was to use national demographics to guarantee representativeness. Then, a screening question was used that excluded individuals who did not have a smartphone. Limiting the data-collection process to those respondents who had a smartphone implies that, in theory, the respondents all had a chance to use a mobile device in the shopping process—an important condition for our research purposes. In simple terms, we started with the ideal of a representative sample but allowed and also preferred the natural "bias" resulting from the fact smartphone ownership might not follow the national demographics.

The online survey yielded 3,160 completed questionnaires, of which 1,027 came from Japan, 1,042 from the USA, and 1,091 from Finland. After careful purification of the dataset based on exclusion of dubious response patterns (e.g., lack of variation between sets of questions), the dataset covered 2,466 respondents. The country-specific distribution for usable data is this: Japan 832 (33.7 %), the USA 776 (31.5 %), and Finland 858 (34.8 %). The basic demographics are depicted in Table 9.1.

3.2 Channel Usage: Offline, Online, and Mobile

We defined the three channels for our respondents thus: "store" refers to a physical shopping environment (department stores, shopping malls, etc.), "online" refers to a Web site that is accessed via a computer (desktop or laptop), and "mobile" refers to the use of software or applications that run on a smartphone or tablet computer. Table 9.2 shows the use of offline, online, and mobile channels in the three main stages of the shopping experience: prepurchase, purchase, and post-purchase.

Table 9.1 Survey demographics

	USA	Japan	Finland	Total
Gender				
Male (%)	49.1	53.0	59.8	54.1
Female (%)	50.9	47.0	40.2	45.9
Age band				
15–24 (%)	28.4	17.8	20.4	22.0
25–34 (%)	25.6	32.3	19.3	25.7
35–44 (%)	22.0	18.6	21.7	20.8
45–54 (%)	16.1	15.4	19.2	17.0
55–70 (%)	7.9	15.9	19.3	14.6
N	776	832	858	2,466

Table 9.2 Channel use in the prepurchase, purchase, and post-purchase stages of shopping in Japan, the USA, and Finland

		USA	Japan	Finland
1. I often visit **stores** for the prepurchase information search (e.g., to find information about retailers, products, and prices)	Avg.	5.9	5.3	5.0
	≥6	54.8 %	44.7 %	42.1 %
2. I often go **online** for the prepurchase information search (e.g., to find information about retailers, products, and prices)	Avg.	7.9	7.1	8.0
	≥6	85.7 %	77.2 %	87.5 %
3. I often use my **mobile** device for the prepurchase information-search phase (e.g., to find information about retailers, products, and prices)	Avg.	6.2	5.9	4.7
	≥6	61.9 %	57.6 %	42.1 %
4. I often visit **stores** to make the purchase (e.g., to pick up, order, or pay for the product)	Avg.	7.7	5.8	7.9
	≥6	85.8 %	53.8 %	85.0 %
5. I often go **online** to make the purchase (e.g., to order or pay for the product)	Avg.	7.2	6.5	6.4
	≥6	80.2 %	68.6 %	69.8 %
6. I often use my **mobile** device to make the purchase (e.g., to order or pay for the product)	Avg.	5.0	5.2	2.8
	≥6	45.1 %	44.5 %	12.9 %
7. I often visit **stores** for the post-purchase information search (e.g., to find instructions for use or address product return issues)	Avg.	5.3	4.3	3.4
	≥6	46.8 %	30.0 %	17.8 %
8. I often go **online** for the post-purchase information search (e.g., to find instructions for use or address product return issues)	Avg.	7.1	5.8	7.3
	≥6	75.8 %	55.0 %	78.3 %
9. I often use my **mobile** device for the post-purchase information search (e.g., to find instructions for use or address product return issues)	Avg.	5.6	5.0	3.9
	≥6	54.4 %	42.9 %	31.0 %

Note: The respondents answered on a 1–10 Likert scale where 1 indicates completely disagreeing and 10 indicates fully agreeing with the statement. For each statement, the upper row shows the means and the lower row the percentage of respondents who agreed with the statement

Several conclusions can be drawn from the data in Table 9.2. Firstly, the data show a tendency toward multichannel shopping among smartphone users. Although the offline channel has a strong role in the purchase stage, the online one was preferred in the pre- and post-purchase stages, a tendency that is likely only to increase with time. Moreover, the role of the mobile channel seems to follow the developments in online usage patterns, although it is still in the development phase. It is also evident that country-specific differences exist. The shoppers in the USA seemed to show versatile usage of all three channels, spanning the prepurchase, purchase, and post-purchase stages of shopping. In comparison by country, the US consumers seemed to use the mobile channel the most. Finland, in contrast, showed the least multichannel behavior. Although Finland was at the forefront in mobile penetration rates, commercial applications for shopping are scarce. This might explain Finland's low percentages for the mobile channel in this study. Finally, Japanese consumers seemed to prefer the online channel, perhaps even to the detriment of the offline channel. The Japanese are also well on the mobile bandwagon, showing strong usage rates in all stages of the shopping experience.

3.3 Use of the Mobile Channel to Enhance Information-Based Value Creation

The mobile channel has a key role in enhancing information-based value creation across channels. To gain better understanding of how the mobile channel can be used during the shopping experience, we identified 18 behaviors and investigated how our respondents viewed them. The results are presented in Table 9.3, where the percentage figures represent respondents' "Yes" answers to the question "Mobile devices or applications can be used in a variety of ways for shopping-related purposes (see the list below). Which of these have you used?" As we saw already in Table 9.2, the differences among the three countries are clear. This holds true for Table 9.3 also: the US consumers had the most experience in using mobile devices for shopping purposes (average: 9.1), followed by the Japanese consumers (average: 7.9). On average, Finnish respondents used 4.6 of the 18 mobile functions.

Table 9.3 The use of mobile devices for shopping-related purposes in the USA, Japan, and Finland

	USA (%)	Japan (%)	Finland (%)
1. Locating a store	88.8	87.5	75.9
2. Using a search engine (e.g., Google) on a mobile device for price and/or product information	83.1	75.5	80.1
3. Checking the availability of a product	67.5	70.1	44.8
4. Creating shopping lists	58.6	28.1	30.8
5. Using a retailer's application for price and/or product information	58.6	47.4	26.7
6. Scanning advertisements (e.g., QR codes or barcodes) for price and/or product information	55.5	55.3	18.5
7. Searching for usage instructions	51.7	59.7	43.6
8. Scanning products (e.g., QR codes or barcodes) while in-store for price and/or product comparisons	50.9	29.0	6.9
9. Redeeming mobile coupons	50.4	66.7	21.3
10. Sharing your shopping experience through social media	45.0	23.0	19.1
11. Giving feedback or making claims/complaints	44.5	36.2	30.1
12. Scanning QR codes or barcodes of the purchased products at home for additional information	44.5	28.8	8.7
13. Making mobile payments via "electronic money" or by credit card	43.4	41.9	19.8
14. Using retailer applications for recreational purposes	38.3	32.1	13.2
15. Collecting customer loyalty points via a mobile application	34.9	38.5	4.8
16. Locating products in the store	33.4	31.5	4.1
17. Logging in with a retailer's application when arriving at a store	33.1	25.8	6.3
18. Saving mobile receipts	30.8	17.8	6.9

Note: The percentage points describe "Yes" responses to the item "Mobile devices or applications can be used in a variety of ways for shopping-related purposes (see the list below). Which of these have you used? Please use "Yes" and "No" to answer"

General functions such as locating a store and using a mobile version of a search engine for price and/or product information were widely used by the respondents in all three countries. While Table 9.3 shows these behaviors from, in general, the most common to the rarest as indicated in our results (sorted by the USA), we will now go further, next providing a thematic classification based on what kind of role these mobile applications and functions have in creation of information-based value in the individual stages of the customer experience.

3.3.1 The Prepurchase Stage

Comparison tools are designed to create economic and functional value for customers by offering tools for sense-making amidst the ever-increasing amounts of price and product information. *Using a search engine (e.g., Google) on a mobile device for price and/or product information* is very common among smartphone users. It is also an example of a third-party influence on the shopping decision-making process.

Using a retailer's application for price and/or product information is about, as the name implies, retailer-specific app use. Imagination is the limit with these applications. For example, an Ikea application serves shoppers with concrete information about dimensions and weights, providing crucial information for evaluating the fit in the house and, for many Ikea customers, in the trailer they will use to carry products home.

Inspiration tools are about meeting information needs related to emotional and symbolic value creation. *Using retailer applications for recreational purposes* is characterized by the ability to serve users with content that is a source of inspiration and also entertainment. Besides Ikea, examples include many fashion retailers, such as Urban Art Guide by Adidas or Amble by Louis Vuitton, which provide interactive content (art and a travel diary, respectively) and allow the user to follow passively or contribute by creating content. Some retailers, among them Lego, have launched mobile games for their customers (Lego App4+ and Lego Creationary). Lego also connects the playful experience to learning and symbolic meanings fostered in the Lego community.

Scanning advertisements (e.g., QR codes or barcodes) for price and/or product information represents the increased interactivity seen in advertising. The camera function of most mobile devices enables applications that can scan various kinds of codes. After scanning a code, the shopper is typically directed to a mobile-device-optimized Web page where additional information is provided. Recently, third-party players such as Blippar have introduced techniques wherein even the codes are not necessary—the interactive content for advertisements is based on picture recognition.

Planning tools serve information that facilitates functional value creation. *Locating a store* is the first step in use of the mobile channel for aiding shoppers with information. Many mobile-environment-optimized Web pages and also purpose-specific applications enable customers to check the location of the nearest store. Some applications also guide the shopper to the store with the aid of GPS data. Examples are abundant and include many retailers and shopping center operators (e.g., Simon Malls), along with third-party service providers and platforms such as Google Maps.

Checking the availability of a product gives the shopper real-time information or an estimate of the availability of a specific product (in a specific store). Also, estimates of delivery times can be provided. Online stores pioneered this function, and many brick-and-mortar retailers have followed.

Creating shopping lists is afforded by tools that simplify choosing and ordering of products. Especially in the context of grocery shopping, the customer's shopping history and loyalty-card data can be utilized for creation of shopping lists that can learn from shopper preferences. Typically, shopping lists can also be shared among, for instance, the members of a family. A Finnish company called Digital Foodie has developed a technology (and application, Foodie.fm) that integrates recipes, store-specific product selections, and a shopping-list function. Furthermore, shopping-list applications have been expanded into tools for customized solutions. For instance, Ikea has developed design tools that assist with such activities as planning a kitchen on the basis of the home's space requirements. In the cross-channel spirit, these plans can also be uploaded to the Ikea cloud, to be opened later in-store by Ikea personnel for further face-to-face consultation.

3.3.2   The Purchase Stage

In-store shopping tools have potential for value creation in both utilitarian and hedonic realms of customer information use. *"Logging in" with a retailer's application when arriving at a store* is a clear example of cross-channel behavior. Customers who enter a brick-and-mortar store sign in with their mobile device by opening the application. Shopkick is one of the pioneers in this field, with an aim of increasing foot traffic in offline stores by rewarding customers who walk in and log in with the app. Besides monetary benefits, the idea of "logging in" is often to serve customers with information about promotions, new products, and perhaps even recreation while one is shopping.

Locating products in the store can use navigation tools for shoppers' in-store use. There are many ways to provide customers with indoor navigation. Most commonly, GPS data or a closed wireless store network is used for determining the customer's location. Some retailers have also experimented with "personal shopper" technologies, as in the case of Emart's Smart Cart service, co-developed with SK Telecom. Smart Cart consists of a mobile application and a physical shopping cart with a Wi-Fi connection and a screen. Shoppers synch their mobile devices (e.g., shopping lists) with the cart and receive navigation help and promotional information.

Scanning products (e.g., QR codes or barcodes) while in-store for price and/or product comparisons is about giving customers tools for comparison and deepening the product information while they are in the store. Best Buy, for instance, has used QR codes in this manner for such products as home theater systems. Many third-party companies, such as RedLaser, have launched their own applications, allowing customers to check whether, for example, the same product retails at a lower price at a nearby store or online. Perhaps a more creative type of solution is represented by Hointer's application: the customer scans a pair of jeans and selects the size desired. In 30 s, the pair of jeans is ready for trying on in a fitting room.

Transaction tools focus on management of shopping-related tasks and information that contribute to functional value. *Making mobile payments via "electronic money" or by credit card* can be facilitated through customer loyalty points or via a third-party transaction. In the first case, the retailer can provide the service, while in the latter case financial institutions and services such as PayPal or the Japanese Suica are used as a platform. QThru is a mobile application that can be used for scanning one's purchases with a mobile device and then making the payment at a self-service kiosk.

Redeeming mobile coupons gives shoppers benefits such as discounts or even free products. A mobile coupon can be redeemed upon showing of a digital coupon that the shopper received by e-mail or downloaded. Coupons can be granted by third-party entities such as coupons.com or the providing companies themselves.

Saving mobile receipts is a natural counterpart to its paper equivalent although not limited to this role. Mobile receipt makes sense especially with products that come with warranties. Besides many retailers, such as Nordstrom, who offer mobile receipts by e-mail, there are third-party services such as Expensify for saving and managing mobile receipts.

Collecting customer loyalty points via a mobile application moves loyalty programs from cards to mobile devices. Tesco has been a pioneer in loyalty programs, and they make no exception when it comes to mobile apps. Tesco's example integrates loyalty data into several shopping tools, such as intelligent shopping lists.

3.3.3 The Post-purchase Stage

Use-value tools are based on information utilized to support customers in—as the name implies—better using their purchased products. The focus, then, is on functional value. *Searching for usage instructions* takes place typically in the post-purchase stage. Online forums and communities have provided resources based on C2C interactions, but recently many retailers too have recognized the value of support for customers after they have left the store. The instructions may be provided by e-mail or via mobile apps or Web pages. Walgreens, for instance, provides a pill reminder for its pharmacy customers, and food brands such as Kraft supply extensive information, including recipes, nutrition details, and video instructions for cooking (Kraft iFood Assistant).

Scanning QR codes or barcodes of the purchased products at home for additional information follows the same idea as in-store scanning of QR codes. In this case, the emphasis is on use-related information.

Communication tools have both a functional and an emotional-symbolic role in information-based value creation. *Sharing your shopping experience through social media* proceeds from shoppers' desire for self-expression. By using social-media interfaces such as Pinterest, Facebook, YouTube, and WhatsApp, shoppers may just post a comment, photo, or video portraying their new possessions. And retailers are facilitating these behaviors. For instance, the Japanese fashion retailer Uniqlo launched Uniqlooks for customers who want to take a photo of themselves wearing Uniqlo clothing and uploaded it so that the community can rate it and celebrate the best styles.

Giving feedback or making claims/complaints is a common feature in many retail apps. This may take place obviously through a traditional telephone service or by e-mail, but chat functions such as Ikea's iconic "Anna" are common too.

4   Discussion and Conclusions

In the age of information, the creation of value for shoppers is in the "choice engines" that support rationalization and inspiration both. With this article, we have striven to illustrate how the need for these engines manifests itself in present shopper behaviors and what kinds of challenges and opportunities they present for the service systems of the future. In doing so, we have emphasized a holistic perspective on customer experience in general and on the shopping experience in particular. As Norman (2009, 52) notes, "In reality, a product is all about the experience. It is about discovery, purchase, anticipation, opening the package, the very first usage. It is also about continued usage, learning, the need for assistance, updating, maintenance, supplies, and eventual renewal in the form of disposal or exchange." It is only through attention to three main temporal perspectives—prepurchase, purchase, and post-purchase—that the need for information and its value and potential can be revealed.

Table 9.4 provides a framework for information-based value creation, summarizing our conclusions. The framework addresses the prepurchase, purchase, and post-purchase stages in terms of three distinct elements:

- The channel-bridging tools for information-based value creation
- Implications for customer value propositions
- Challenges and opportunities for service system development

In the prepurchase stage, customers' information needs can be served with comparison tools, inspiration tools, and planning tools. These tools point to implications for economic, functional, emotional, and symbolic dimensions of customer value propositions alike. From the perspective of service system development, the key questions revolve around opening up organizational product and price data for customers and providing/enhancing interfaces between C2C and third-party platform actors.

Table 9.4 The framework for information-based value creation

	Prepurchase	Purchase	Post-purchase
Channel-bridging tools for information-based value creation	**Comparison tools** • Using a search engine (e.g., Google) on a mobile device for price/product information • Using a retailer's application for price and/or product information **Inspiration tools** • Using a retailer's application for recreational purposes • Scanning advertisements (e.g., QR codes or barcodes) for price and/or product information **Planning tools** • Locating a store • Checking the availability of the product • Creating shopping lists	**In-store shopping tools** • "Logging in" for special deals and recreation • Locating products and navigating the store • Scanning products for additional information and comparisons **Transaction tools** • Making mobile payments • Redeeming mobile coupons • Saving mobile receipts • Collecting customer loyalty points via a mobile application	**Use-value tools** • Utilizing mobile usage instructions • Finding additional product information through code-scanning **Communication tools** • Sharing the shopping experience • Submitting feedback/complaints

(continued)

Table 9.4 (continued)

	Prepurchase	Purchase	Post-purchase
Implications for customer value propositions	**Economic value:** Demonstrating the best monetary value (absolute price or total cost) and gaining price transparency **Functional value:** Finding superior solutions for matching products and learning from customer preferences through search and comparison patterns **Emotional value:** Finding tools and interactive content for inspiration and entertainment **Symbolic value:** Benefiting from interactive content for fostering of the brand relationship	**Economic value:** Real-time and location-specific discounts and special offers, along with dynamic pricing **Functional value:** Navigation tools, additional product and price information, shopping lists, and adaptive solutions for payment and ordering **Emotional value:** Enriching the shopping experience with digital information for entertainment and stimulation **Symbolic value:** Increasing the visibility of shopping trip and related purchase decisions through social media	**Economic value:** - **Functional value:** Supporting the user experience and providing a convenient after-sales experience **Emotional value:** Extending the shopping experience and creating peace of mind **Symbolic value:** Facilitating customers' self-expression
Challenges and opportunities for service system development	• Utilizing the data in company information systems for serving customers with products, availability, and price information • Enriching retailer-specific product information with C2C information (e.g., via social media) • Increasing or restricting information from third-party platforms	• In-store technologies that utilize the data in company information systems for serving customers and personnel with product-, availability-, and price information • Cloud-based (or wireless) technologies that can identify customers and utilize their prior purchase patterns • Arming personnel with shoppers' digital prepurchase-phase data	• Utilizing the data in company information systems for serving customers with a wide range of product information, such as use instructions and technical details • Facilitating C2C interactions for peer support and fostering virtual customer communities • Integrating the help desk for multiple channels

The purchase stage employs two kinds of tools: in-store shopping tools and transaction tools. These tools may also be used in development of more resonating and coherent customer value propositions for all four dimensions. Challenges and opportunities for service system development arise especially in relation to interfaces between in-store technologies, responsive app use in-store, and tools for service personnel.

Finally, the post-purchase stage features use-value tools and communication tools. These can be utilized for enhancing functional, emotional, and symbolic customer value propositions. The challenges for service system development in the post-purchase stage are related to product-specific information that supports the assembly and use of the products purchased. Other considerations include C2C interactions and new ways to provide help-desk services.

For retailers, our exploratory results and conceptual findings suggest that:

- Many customers are already utilizing multiple channels for comparing products and service providers—hence, retailers that actively provide tools for supporting customers' choice processes and meeting other information needs are better equipped in the otherwise commoditized marketplace, where loyalty is easily supplanted by the lowest price.
- Customer experience management facilitated by relevant information tools is becoming the new CRM. The value is not in the exponential growth of information but in helping customers to make sense of it and enriching their life. Since customers truly derive benefits from informational tools, they are more willing to share their purchase histories and contextual preferences with retailers. In the near future, this may make some of the current loyalty-card systems obsolete.
- Information value creation should support the chosen strategic customer value propositions, whether they are focused around price (economic value), solutions (functional value), customer experience (emotional value), meanings (symbolic value), or some combination of these.
- The developments in tools and models of sharing information raise issues of data privacy and trust once again as a potential concern for consumers.
- While the mobile channel is the key to bridging the offline and online worlds, employees currently lack tools to serve the emerging cross-channel customers.
- All the aforementioned issues present challenges along with their possibilities for the design, integration, and management of various service systems.

We believe information-based value creation is a theme that warrants further research. Our research being exploratory and conceptual in nature, future endeavors might benefit from development of metrics for information value in the prepurchase, purchase, and post-purchase stages of the shopping experience. It would then be possible to relate these measurements to the study of service systems, thereby contributing to the evaluation and development of future service systems. Moreover, the theme and logic of informational value can be expanded to other areas of life than shopping. What are the informational needs of other types of service business? How could we as citizens fare better as we navigate the complex systems of, for example, health care, taxation, and education?

References

Anderson JC, Narus JA, Van Rossum W (2006) Customer value propositions in business markets. Harv Bus Rev 84(3):91–99

Archer-Brown C, Piercy N, Joinson A (2013) Examining the information value of virtual communities: factual versus opinion-based message content. J Market Manag 29(3–4):421–438

Beath C, Berecca-Fernandez I, Ross J, Short J (2012) Finding value in the information explosion. MIT Sloan Manag Rev 53(4):18–20

Chatterjee P (2010) Multiple-channel and cross-channel shopping behavior. Role of consumer shopping orientations. Market Intell Plann 28(1):9–24

Cho M-H, Jang S (2008) Information value structure for vacation travel. J Trav Res 47(1):72–83

Day GS, Moorman C (2010) Strategy from the outside in: profiting from customer value. McGraw-Hill, New York

Deloitte (2012) Switching channels: global powers of retailing 2012. Accessed 1 Feb 2012. http://www.deloitte.com/assets/Dcom-Mexico/Local%20Assets/Documents/mx%28en-mx%29Global_Power_Retailing.pdf

Demirkan H, Spohrer JC, Krishna V (2011) Introduction of the science of service systems. In: Hefley B, Murphy W (eds) Service science: research and innovations in the service economy. Springer, New York

Hirshleifer J (1973) Where are we in the theory of information? Am Econ Rev 63(2):31–39

Larivière B, Joosten H, Malthouse EC, van Birgelen M, Aksoy P, Kunz WH, Huang M-H (2013) Value fusion: the blending of consumer and firm value in the distinct context of mobile technologies and social media. J Serv Manag 24(3):268–293

Lusch RF, Vargo SL, O'Brien M (2007) Competing through service: insights from service-dominant logic. J Retailing 83(1):5–18

Maglio PP, Spohrer J (2008) Fundamentals of service science. J Acad Market Sci 36(1):18–20

Maglio PP, Vargo SL, Caswell N, Spohrer J (2009) The service system is the basic abstraction of service science. Inform Syst E Bus Manag 7:395–406

Mele C, Polese F (2011) Key dimensions of service systems in value-creating networks. In: Hefley B, Murphy W (eds) Service science: research and innovations in the service economy. Springer, New York

Norman DA (2009) Systems thinking: a product is more than the product. Interactions 16(5):52–54

Novani S, Kijima K (2012) Value co-creation by customer-to-customer communication: social media and face-to-face for case of airline service selection. J Serv Sci Manag 5:101–109

Payne A, Frow P (2014) Developing superior value propositions: a strategic marketing imperative. J Serv Manag 25(2):213–227

Peterson M, Gröne F, Kammer K, Kircheneder J (2010) Multi-channel customer management: delighting consumers, driving efficiency. J Direct Data Digit Market Pract 12(1):10–15

Resmini A, Rosati L (2011) Pervasive information architecture: designing cross-channel user experiences. Morgan Kaufmann, Burlington

Rigby D (2011) The future of shopping. Harv Bus Rev 89(12):64–75

Rintamäki T, Kuusela H, Mitronen L (2007) Identifying competitive customer value propositions in retailing. Manag Serv Qual 17(6):621–634

Spohrer J, Kwan SK (2009) Service science, management, engineering and design (SSMED): an emerging discipline—outline & references. Int J Inform Syst Serv Sector 1(3):1–31

Spohrer J, Maglio PP (2010) Service science: towards a smarter planet. In: Karwowski W, Salvendy G (eds) Introduction to service engineering. Wiley, New York, pp 3–30

Thaler RH, Tucker W (2013) Smarter information, smarter consumers. Harv Bus Rev 91(1):45–54

Vargo SL, Lusch RF (2004) Evolving to a new dominant logic of marketing. J Market 68(1):1–17

Vargo SL, Lusch RF (2008) Service-dominant logic: continuing the evolution. J Acad Market Sci 36(2):1–10

Webster FE Jr (1994) Market-driven management. Using the new marketing concept to create a customer-oriented company. Wiley, New York

Zhang J, Farris PW, Irvin JW, Kushwaha T, Steenburgh TJ, Weitz BA (2010) Crafting integrated multichannel retailing strategies. J Interact Market 24(2):168–180

Chapter 10
Service R&D Program Design Aiming at Service Innovation

Yuriko Sawatani and Yuko Fujigaki

Abstract The society has been fast advancing toward a service-based economy. This phenomenon, common to both developed and developing countries, results from the growth of the service sector's share of the economy, spurred by rapid growth in service industries consequent to increased social sophistication and diversification. This affects the research and development (R&D) organization, so R&D outcomes are expected to contribute to service innovation. Based on these phenomena, a program concept is introduced to the government-funded R&D to strengthen the linkage between R&D and innovation. In addition, service R&D has been focused triggered by service science initiatives. This chapter discusses service R&D program design for service innovation. Most of the design activities are done at the planning phase. However, the execution-phase activities are more important to achieve program-level objectives by strengthening the linkage between R&D and innovation. These interactions between a program and projects create values that are not expected at the planning phase, so we should have a program management to encourage these post-value co-creation characteristics.

Keywords Information technology • R&D management • Service innovation • Service science

1 Introduction

The structural change in society of the shift to a service-based economy is advancing. This phenomenon, common to the economies and societies of both developed and developing countries, results from the growth of the share of the economy

An earlier version of this paper was presented at the Human Side of Service Engineering (HSSE) in 2012 (Sawatani and Arimoto 2012).

Y. Sawatani (✉)
Center for Research Strategy, Waseda University, Tokyo, Japan
e-mail: yurikosw@gmail.com

Y. Fujigaki
Department of General Systems Studies, The University of Tokyo, Tokyo, Japan

accounted for by the service sector, spurred by rapid growth in service industries against a background of increasing social sophistication and diversification. The service industry's share of Japanese GNP has grown to 60.7 % (the percentage of nominal GDP in fiscal 2009 not including the agriculture, forestry, fisheries, mining, and manufacturing industries), and according to a study by the Organisation for Economic Co-operation and Development (OECD), the shift to a service-based economy has advanced steadily in other countries too. However, productivity in service industries is lower than in manufacturing industries, and the need to achieve innovation and productivity improvements has become an important issue (Ministry of Economy and Trade and Industry of Japan (METI) 2007).

Regarding this issue, the report "Innovate America: Thriving in a World of Challenge and Change" submitted to the Bush administration by the US Council on Competitiveness in December 2004 (commonly referred to as the "Palmisano Report") offered the view that there was a need to create an interdisciplinary field of service science to resolve issues originating in the shift to a service-based economy (IfM and IBM 2007; Chesbrough and Spohrer 2006). A new movement has appeared toward the development of such an interdisciplinary field of service science as science and engineering researchers join the domain of service research that until now has advanced chiefly in the fields of social science, service marketing, and service management. In Japan, addressing emerging and interdisciplinary domains over the years 2006 through 2010 was planned in March 2006 under the Third Science and Technology Basic Plan. In the Fourth Science and Technology Basic Plan, the focal point has shifted further from field-specific to issue-driven innovation in science and technology, with research and development activities in interdisciplinary domains such as service science positioned as important topics and serving as forerunners of the issue-driven approach. The June 2006 outline of the Economic Growth Strategy from the Ministry of Economy, Trade and Industry discussed innovation in service industries, and a movement to create a field of service science has begun in Japan too. In May 2007 Service Productivity & Innovation for Growth (SPRING) was established, as was the Center for Service Research (at the National Institute of Advanced Industrial Science and Technology, or AIST) in April 2008; in April 2007 the Ministry of Education, Culture, Sports, Science and Technology began the Service Innovation Human Resource Development Program; and in April 2010 the Japan Science and Technology Agency's Research Institute of Science and Technology for Society (JST-RISTEX) began seeking R&D projects under its Service Science, Solutions and Foundation Integrated Research Program (S3FIRE).

At the same time, the definition of services has been reconsidered. In industry categories, the service industry refers to what is left over after grouping into the agriculture, forestry, fisheries, and manufacturing industries. In service marketing, an attempt was made to separate products and services and define services using characteristics differing from those of products. In recent years, service-dominant (S-D) logic has been proposed, seeing the essence of services as the co-creation of value and identifying service as a fundamental principle of exchange, and research based on this concept is in the process of spreading not just in service marketing but to other fields as well. The following section reviews service R&D-related studies and discusses service R&D model focusing on this new paradigm.

2   Existing Literatures on Service R&D and Service Innovation

2.1   Service R&D

R&D definition has been changed and influenced by economic changes and servitization. The definition of R&D has been worked in National Experts on Science and Technology Indicators (NESTI) since 1960 and updated in Frascati Manual version 6 (OECD 2002). Originally R&D focused on natural science and engineering; however, the wider R&D definition including social sciences and humanities was considered by Djellal et al. (2003) when Frascati Manual version 6 was created:

> Research and experimental development (R&D)
>   Research and experimental development (R&D) comprise creative work undertaken on a systematic basis in order to increase the stock of knowledge, including knowledge of man, culture and society, and the use of this stock of knowledge to devise new applications. (OECD 2002 p. 20)

The R&D definition includes "knowledge of man, culture and society" and becomes wider from technological outputs to social science-based knowledge creation. The R&D in social sciences and humanities is described as the following:

For the social sciences and humanities, an appreciable element of novelty or a resolution of scientific/technological uncertainty is again a useful criterion for defining the boundary between R&D and related (routine) scientific activities. This element may be related to the conceptual, methodological or empirical part of the project concerned. Related activities of a routine nature can only be included in R&D if they are undertaken as an integral part of a specific research project or undertaken for the benefit of a specific research project. Therefore, projects of a routine nature, in which social scientists bring established methodologies, principles and models of the social sciences to bear on a particular problem, cannot be classified as research." (OECD 2002 p. 48)

"Defining the boundaries of R&D in service activities is difficult, for two main reasons: first, it is difficult to identify projects involving R&D; and, second, the line between R&D and other innovative activities which are not R&D is a tenuous one. ….Identifying R&D is more difficult in service activities than in manufacturing because it is not necessarily "specialised". It covers several areas: technology-related R&D, R&D in the social sciences and humanities, including R&D relating to the knowledge of behaviour and organisations. ….

Also, in service companies, R&D is not always organised as formally as in manufacturing companies (i.e. with a dedicated R&D department, researchers or research engineers identified as such in the establishment's personnel list, etc.). The concept of R&D in services is still less specific and sometimes goes unrecognised by the enterprises involved. As more experience becomes available on surveying R&D in services, the criteria for identifying R&D and examples of service-related R&D may require further development. ….

"The following are among the criteria that can help to identify the presence of R&D in service activities:

- Links with public research laboratories.
- The involvement of staff with PhDs, or PhD students.
- The publication of research findings in scientific journals, organisation of scientific conferences or involvement in scientific reviews.
- The construction of prototypes or pilot plants (subject to the reservations noted in Sect. 2.3.4)." (OECD 2002 p. 48–49)

Even though some guidelines determine whether routine works or service R&D are suggested, those are not enough to describe how to design service R&D.

A new knowledge production concept focusing on issue-based research was proposed as a mode 2 (Gibbons et al. 1994). Mode 2 knowledge production in an application context, which includes the experiential elements as well as theoretical elements, has the similar characteristics as the service R&D activities from the viewpoints of the participation of various stakeholders and the quality control by them. However, as the presented framework is confined to the general idea to describe the characteristics of the knowledge creation with transdisciplinary, the positive research that aimed at the elucidation of the R&D behavior is not being done fully.

2.2 Service Innovation

The macro-level innovation surveys are conducted including service industries. Miles pointed out that there are two issues on service innovation surveys (Miles 2002, 2007). One is the survey design that is biased to the technological innovation. The current survey questionnaires depend on the innovation studies based on goods innovation (Miles 2007; Drejer 2004) and could not capture the wider scope of service innovation. The other issue is the immature understanding of service innovation by service industries in particular (Miles 2007; Sundbo 1997). Service industries do not have a specialized innovation organization such as R&D in most cases (Miles 2007; Sundbo 1997), so it is difficult for them to recognize activities and knowledge contributing to the service innovation.

On the other hand, case studies of service innovation identified non-technological innovation, such as process and organizational innovation adding to the technological innovation (Miles 2002; Sundbo 1997). The empirical findings of the service innovation show that the characteristics of the service innovation processes are dynamic and ad hoc (Sundbo 1997; Mamede 2002; Edvardsson and Olsson 1996). Despite the growing studies on service innovation, the literature from marketing and innovation research continues to improve our understanding of service innovation; however, these service innovation studies show the tendency to emphasize the service distinctive features (Gallouj and Weinstein 1997; Gallouj 2002; Sundbo 1997; de Vries 2006; van der Aa and Elfring 2002) or the assimilation of goods and services (Gallouj 1998; Vargo and Lusch 2004a, b; Drejer 2004). Both approaches are based on the separation of goods or services and provide incomplete view of service innovation (Drejer 2004). The modern service theory is being formed based on value co-creation by customers and service providers (Vargo and Lusch 2004a, b; Vargo et al. 2008, 2010).

2.3   Service R&D Model

The S-D logic has been recognized as the theoretical foundation of service research (Vargo and Lusch 2004a, b). It expects to integrate the traditional G-D logic view of innovation model based on goods vs. services dichotomy to a unified innovation model from the value co-creation point of views. The S-D logic is appropriate for studying service innovation since it removes the limitation of goods vs. intangible goods (services) dichotomy approaches and synthesizes customers and service providers (Drejer 2004).

Moeller (2008) shows service processes based on collaboration between service providers and customers influenced by S-D logic. The service processes are divided into facilities, transformation, and usage. The stage of facilities exhibits only when potential value for customers is created by company resources. The next stage, transformation, is divided into two parts: company-induced transformation or customer-induced transformation. At the company-induced transformation, the transformation includes only company resources for the potential value for customers. On the other hand, at the customer-induced transformation, companies and customers co-create value. At the last stage, the usage, customers act as the prime resource integrator to receive benefits from the transformation. The facilities and the company-induced transformation are both activities of service providers. So the Moeller's model has three types of activities, such as activities done by service providers alone (facilities and company-induced transformation), value co-creation activities by both (customer-induced transformation), and activities of customers (usage).

We introduce service R&D model extending the Moeller's model (Sawatani and Fujigaki 2014). The model has three spheres, such as R&D, value co-creation, and site. "R&D activities" and "new research theme creation" are activities contained in a R&D sphere (Sawatani and Niwa 2008). There are bidirectional links between a R&D and a value co-creation sphere. The first link from "R&D activities" to "value co-creation" implies that a R&D sphere provides technologies and knowledge to a value co-creation sphere. The service innovation success depends on technologies and knowledge created in a R&D sphere integrated by the design methods at a value co-creation sphere (Sawatani and Niwa 2009).

The second link illustrates that a R&D sphere gains research value through the value co-creation interaction with customers and members in a service organization, not only providing their technologies and knowledge to them. The knowledge created through the value co-creation interaction includes technologies, integrated design methods, and service domain knowledge (Sawatani and Niwa 2009). Adding to the knowledge creation, new research themes are discovered when

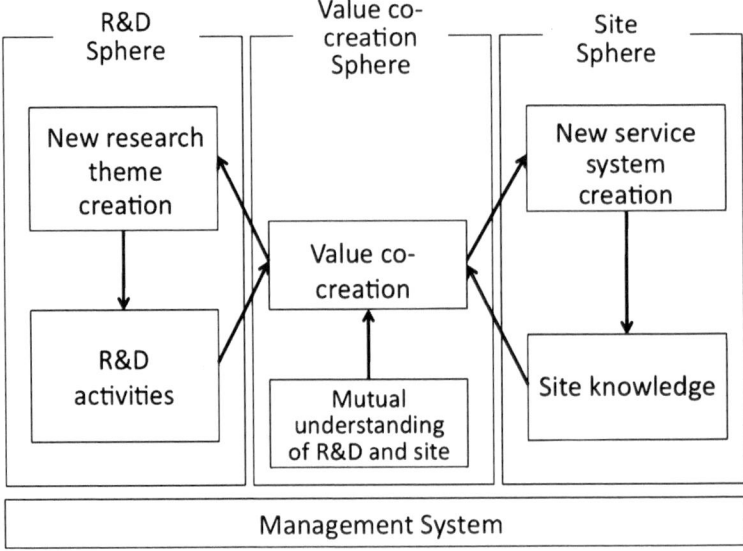

Fig. 10.1 Service R&D model

researchers are practicing activities in a value co-creation sphere (Sawatani and Niwa 2008). The value co-creation interaction is beneficial not only to a site sphere for customers but also to a R&D sphere (Vargo et al. 2010).

In this model, researchers are not only providing technologies and knowledge into a service sphere but also receiving new research ideas by proactively joining to the value co-creation interaction (Sawatani and Niwa 2008, 2009). The S-D logic states that "value created through exchange is based on the mutually beneficial relationships among service systems" and "all parties are simultaneously both producers and customers of value" (Vargo et al. 2010). That is, the value co-creation interaction is beneficial to both the R&D and site spheres. The key element of the service research model is the value co-creation sphere where it requires the collaboration of researchers and customers/service members (Fig. 10.1).

In addition to these R&D and site activities, management system is added to cross these three spheres. This works for a supplementary system to execute R&D projects and will provide vital functions.

3   Service R&D Program Design

The program is a set of projects to produce the program-level outcomes (PMI, BIS, HM Treasury). The government introduced a program concept to R&D program to drive innovation through funding R&D projects. Table 10.1 shows the various R&D projects.

Government-funded strategic R&D program drives selected R&D projects to meet the program-level objectives aiming at innovation. The underlined word(s) in Table 10.1 shows the similarity with government-funded strategic R&D features. Corporate R&D has similar characteristics with government-funded strategic R&D except R&D period and funding source. Next we look into a corporate service R&D program and a government-funded service R&D program in more detail.

3.1   Corporate Service R&D Program

The service R&D projects develop a service system interacting with service receivers. Using two types of outputs from service R&D projects (knowledge base or knowledge-embedded service system) and inputs from service receivers (high intensity or low intensity), we developed a conceptual framework (Sawatani and Niwa 2009) to categorize service projects (Table 10.2). Typical service R&D

Table 10.1   Features of R&D project

	Government-funded strategic R&D	Government-funded curiosity-driven R&D	Corporate R&D	Projects in business IT construction
Created value	Scientific, social, and economic values	Mainly scientific	Scientific, social, and economic values	Social and economic values
Open/closed system	Open system	Mainly closed system	Open system	Mainly closed system (customer's system)
Period	Long term	Long term	Short term	Case by case
R&D organization	Multidisciplinary	Various, mainly single discipline	Mainly single discipline	Mainly single discipline
Public/private fund	Public	Public	Private	Private
Uncertainty	High	High	High	Low

Table 10.2   Service system framework: service project categories

	Intensity of service receivers	
	High	Low
Produced value		
Knowledge base	Professional services (open pattern)	NA
Knowledge-embedded service system	IT-supported front-stage services (interactive pattern)	IT-supported back-stage services (closed pattern)

projects in the high intensity x knowledge base quadrant are professional services, such as R&D management services, and innovation management services which mainly produce knowledge for service receivers. IT-supported front-stage services are customer relationship management (CRM) and business process management (BPM). Optimization projects, such as SCM, are example of projects in the low intensity x knowledge-embedded service system quadrant, which are mainly in the back stage of a service system.

There are several issues on the execution of service R&D projects. Service receivers might not be able to express their problems clearly. On the other hand, researchers do not understand their requests or issues clearly due to a lack of local knowledge of the site. Researchers tend to stick to the current discipline area and do not explore those issues enough from service receiver's viewpoints. To make service R&D projects successful, the service R&D management needs to support researchers to explore service R&D areas. In addition to these common issues, different considerations are necessary depending on the types of service R&D projects.

3.2 Government-Funded Service R&D Program

We look into the Service Science, Solutions and Foundation Integrated Research Program (S3FIRE, RISTEX), one of the government-funded service R&D programs, as a case study. It started since 2010. The objective of the S3FIRE program is to establish a research foundation for service science by developing technologies and methodologies to solve problems effectively and creating a community among researchers and practitioners. There are two types of research approaches as the following:

Type A: Research on solution-development
Type B: Research on SSME scientific foundation

The program aims that complementary function between solution-development A and creation of scientific concepts, theories, technologies, and methodologies B shall establish scientific foundation for SSME. S3FIRE program is designed considering the following points:

- Objectives of the program
- Organization of the program
- Funding types of the program, such as project types and costs
- Program management
- Project selection and monitoring

The program management shows not only how to make projects successful but also how to create program-level outputs interacting with projects. Program management activities at S3FIRE are added to Fig. 10.2.

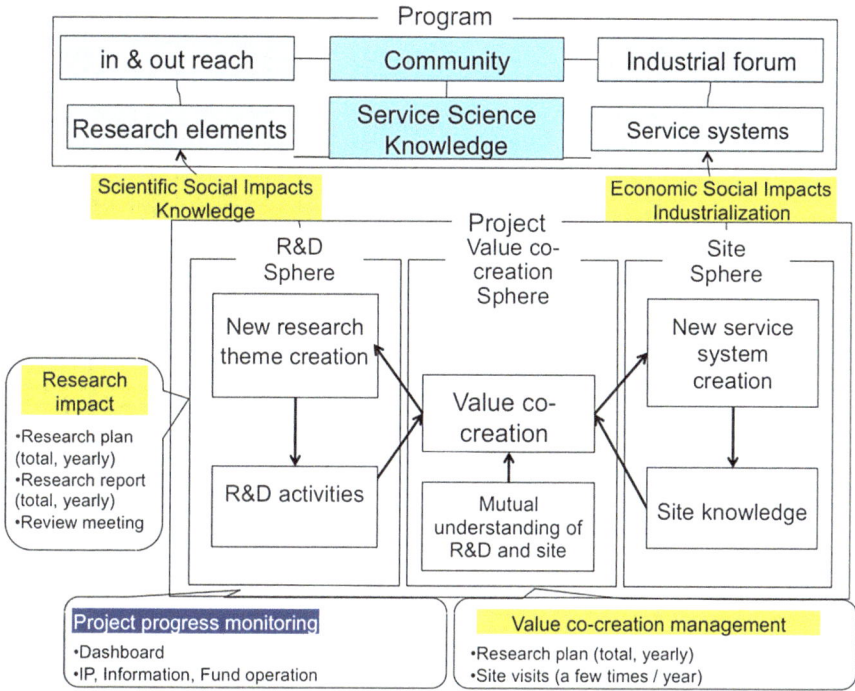

Fig. 10.2 S3FIRE program management

3.3 Design Points of Service R&D Program

Table 10.3 shows the design points of service R&D and related programs (BIS 2010; HM Treasury 2011). The ROAMEF is a reference model of government program, focusing on a program's life cycle, such as rationale, objective, appraisal, monitoring, evaluation, and feedback. Activities at the planning phase are similar for most cases. However, S3FIRE (government-funded service R&D program) and corporate service R&D case include program management activities at the execution phase, not only selecting/monitoring but also interacting with projects. Most of the design activities are done at the planning phase. However, the execution-phase activities are more important to achieve program-level objectives by strengthening the linkage between R&D and innovation. These interactions between a program and projects create values that are not expected at the planning phase, so it is important for the program design to include a mechanism of the program management evolution.

Table 10.3 Service system framework: service project category

Activities	Program with high uncertainty			Program based on social needs	Corporate service R&D	Program with low uncertainty
	Government-funded program					Projects in business IT
	S3FIRE	ROAMEF				
Plan						
Objectives	O	O	Rationale, objective	O	O	O
Organization	O	O		O	O	O
Funding	O	O			O	O
Program management design	O	O	Appraisal, monitoring		O	O
Execution						
Project selection, monitoring	O	O			O	O
Project-level value co-creation management	O				O	O
Program-level value co-creation management	O				O	O
Evaluation, review, assessment						
Ex ante evaluation	O	O	Evaluation		O	O
Midterm evaluation	O	O			O	
Ex post evaluation	O	O				
Feedback		O	Feedback			

4 Toward Service Innovation Through Service R&D Program

This paper discusses service R&D program design for service innovation, referring to the Service Science, Solutions and Integrated Research Program (S3FIRE). The execution-phase activities are more important to achieve program-level objectives by strengthening the linkage between R&D and innovation. The program design needs to evolve during a program period. This causes issues of program evaluation. The program has the characteristics of value co-creation with projects, which happen after the program started (post-value co-creation characteristics). Service R&D programs are open systems, so the current evaluation based on predefined activities at the planning phase has the possibility to limit the outputs and outcomes of the program. To maximize the future impacts of the program, we need to look for a new cooperative program evaluation method (Fig. 10.3).

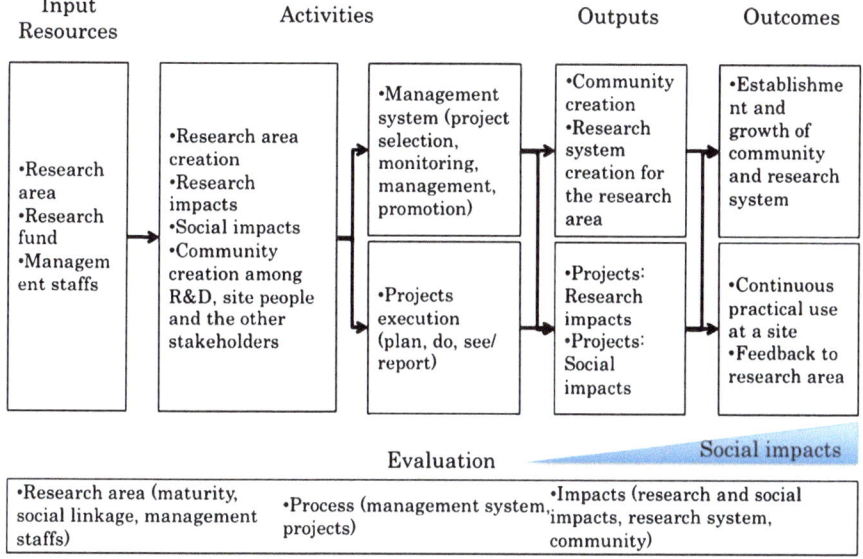

Fig. 10.3 Logic model of service R&D program (S3FIRE)

References

BIS (2010) Guideline for managing programmes. http://www.bis.gov.uk/assets/biscore/corporate/docs/g/10-1256-guidelines-for-programme-management

Chesbrough H, Spohrer J (2006) A research manifesto for services science. Comm ACM 49(7):35–40

de Vries EJ (2006) Innovation in services in networks of organizations and in the distribution of services. Res Pol 35(7):1037–1051

Djellal F, Francoz D, Gallouj C, Gallouj F, Jacquin Y (2003) Revising the definition of research and development in the light of the specificities of services. Sci Publ Pol 30(6):415–429

Drejer I (2004) Identifying innovation in surveys of services: a Schumpeterian perspective. Res Pol 33(3):551–562

Edvardsson B, Olsson J (1996) Key concepts for new service development. Serv Indust J 16(2):140–164

Gallouj F (1998) Innovating in reverse: services and the reverse product cycle. Eur J Innovat Manag 1(3):123–138

Gallouj F (2002) Knowledge-intensive business services: processing knowledge and producing innovation. Edward Elgar, UK

Gallouj F, Weinstein O (1997) Innovation in services. Res Pol 26:537–556

Gibbons M, Limoges C, Nowotny H, Schwartzman S, Scott P, Trow M et al (1994) The new production of knowledge. Sage, London

HM Treasury (2011) THE GREEN BOOK Appraisal and Evaluation in Central Government. http://www.hm-treasury.gov.uk/d/green_book_complete.pdf

IfM IBM (2007) Succeeding through service innovation: a service perspective for education, research, business and government. University of Cambridge Institute for Manufacturing, Cambridge

Mamede R (2002) Does innovation (Really) matter for success? The case of an IT consultancy firm. In: DRUID conference on industrial dynamics of the new and old economy, Elsinore, 6–8 June 2002

Ministry of Economy, Trade and Industry of Japan (METI) (2007) Service Innovation report. http://www.meti.go.jp/report/data/g70502aj.html

Miles I (2002) Service innovation: towards a tertiarization of innovation studies. In: Gadrey J, Gallouj F (eds) Productivity, innovation and knowledge in services. Edward Elgar, Cheltenham

Miles I (2007) Research and development (R&D) beyond manufacturing: the strange case of services R&D. R&D Manag 37:249–268

Moeller S (2008) Customer integration—a key to an implementation perspective of service provision. J Serv Res 11(2):197–210

OECD (2002) Frascati manual—proposed standard practice for surveys on research and experimental development. OECD, Paris

RISTEX (Research Institute of Science and Technology for Society) http://www.ristex.jp/EN/aboutus/principle.html

Sawatani Y, Arimoto T (2012) Value co-creation in R&D. Advances in the Human Side of Service Engineering Edited by Louis E . Freund CRC Press 2012 Pages 37–47

Sawatani Y, Fujigaki Y (2014) Transformation of R&D into a Driver of service innovation: conceptual model and empirical analysis. Informs Serv Sci 6(1):1–14

Sawatani Y, Niwa K (2008) Services research model for value co-creation. Management of Engineering & Technology, PICMET, pp 2354–2360

Sawatani Y, Niwa K (2009) Service systems framework focusing on value creation: case study. Int J Web Eng Tech 5(3):313–326

Sundbo J (1997) Management of innovation in services. Serv Indust J 17(3):432–455

van der Aa W, Elfring T (2002) Realizing innovation in services. Scand J Manag 18:155–171

Vargo SL, Lusch RF (2004a) The four service marketing myths: remnants of a goods-based, manufacturing model. J Serv Res 6(4):324–335

Vargo SL, Lusch RF (2004b) Evolving to a new dominant logic for marketing. J Market 68(1):1–17

Vargo SL, Maglio PP, Akaka MA (2008) On value and value co-creation: a service systems and service logic perspective. Eur Manag J 26:145–152

Vargo SL, Lusch RF, Akaka MA (2010) Advancing service science with service-dominant logic. In: Maglio PP, Kieliszewski CA, Spohrer JC (eds) Handbook of service science, Springer, UK, pp 134–156

Index

A
Access rights, 21
After innovation, 90
Algorithms, 88
Artificial intelligence, 61, 65

B
Bedside to bench, 40
Bench to bedside, 40
Bottom-up management, 89
Boundaries, 43, 76–77
Boundary objects, 56, 57, 62
Business/information architecture, 110
Business model, 109
Business modeling software, 126
Business motivation, 113
Business Motivation Model (BMM), 127

C
Characters, 61
City innovations, 87
Co-creation, 62, 67
Co-definition of shared internal model, 50
Co-development of value, 50
Co-elevation of each other, 50
Co-existence, 106
Co-experience of service, 50
Collaborations, 18
Collaboration with users, 90
Communications and control, 43
Competencies, 77–78
Competitions, 18
Complexity, 43
Complex systems, 99

Conceptual innovation, 87
Co-prosperity, 106
Core service review, 123–124
Cretan's paradox, 103
Cross-channel strategy, 148
Customer
    experience, 158
    service, 62
    value, 149
Customer value propositions, 149, 161
Cybernetics, 43

D
Data model, 111
Dichotomy, 105
Digital platforms, 55, 58, 65
Dissemination, 93

E
Ecology, 15
Economics, 15
Eco-system, 87
Emergence, 74–76
Emotions, 64, 66
Employee-driven innovation, 84
Empowerment, 85
Empowerment strategies of stakeholders, 53
Engagement, 65
Entanglement, 74
Enterprises, 66
Entities, 17
Episteme, 40
Exchange, 99, 12
External sources of innovations, 91

F
Federal Enterprise Architecture (FEA), 112
Flexible front ends, 58, 59
Four-phase Value Co-Creation Process
        Model, 49
Frameworks, 73
Functional model, 111

G
Gap Model of Service Quality, 38
General systems theory, 71
Global simulation, 26
Goods-dominant logic, 40
Governance mechanisms, 20
Government
    program, 115
    service, 116

H
Hard systems approach, 44
Hierarchy, 43
Holism, 73–75

I
Indirect exchange, 13
Information value, 150, 151
Information value creation, 146, 151, 158, 161
Infusion and growth, 23
Interactions, 17
Interdisciplinary approach, 39
Internal model, 44, 46
Involvement strategy, 51–52
ISO certification, 122–123
ISSIP.org, 123

K
Knowledge, 99, 62
Knowledge processes, 57

L
Law of Requisite Variety, 43
Literature, 70
Logic, 97

M
Measures, 21
Measuring and optimizing, 25

Mobile value creation, 151
Multichannel strategy, 148
Mutually exclusive and collectively exhaustive
        (MECE), 98

N
Network externalities, 57
Nordic School Approach, 39

O
OMG Government Domain Task Force, 126
Online banking, 60
Open innovation, 91
Open systems, 43
Organizational/governance reviews, 121
Outcomes, 115
Outside
    in knowledge flows, 92
    in thinking, 147

P
Participatory dynamics, 85
Performance measures, 119
Perspectives, 76–77
Phenomenological, 15
Phronesis, 40
Platform
    leader, 67
    participants, 56
Policy competences, 93
Post-purchase stage, 152, 157–158
Prepurchase stage, 152, 155–156
Principle of Internal Modeling, 44
Public policy, 116
Purchase stage, 156–157

R
Reductionism, 74, 98
Reference model, 109
Relationship, 61
Research and development (R&D), 163
Resource integration, 77–78
Resources, 71

S
Self-referential paradox, 103
Self-service, 60
Service artifacts, 56, 60, 61, 66

Service-based budgeting, 121
Service-Dominant (S-D) Logic, 112, 39, 57, 70–73, 91
Service economies, 14
Service engineering, 44
Service exchange, 58
Service-for-service, 12
Service innovation, 83, 166
Service offerings, 64
Service output, 117
Service R&D, 165, 167
Service R&D program, 168, 171
Service science, 164
Service science management and engineering (SSME), 70
Service system, 39, 72, 146, 148, 161
Service systems science, 37, 42
Service systems theories and models, 42–44
Service transformation, 124–125
Service value, 117
Share and Spread (SIPS), 52
5S-KAIZEN, 129
Social entities, 55
Social graph, world, 30
Social innovations, 84
Social problems, 85
Social values, 45
Societal embedding, 86
Soft systems approach, 44
Stakeholder roles, 21
Strategic plans, 112
Subjectivity, 100
System innovations, 84

T
Target groups, 115
Techne, 40
The Open Group Architecture Framework (TOGAF), 127
Third sector, 86
Top-down principle, 88
Transdiscipline, 44
Translational approach, 45
Translational systems sciences, 40
T-shaped, 88
Two-sided market, 66

U
Uncertainty principle, 99
Unique, 15
User-driven innovation, 84
User needs, 89
Users as innovators, 89

V
Value
    of business models, 47
    co-creation, 55, 39, 56, 58, 72
    curation strategy, 52
    of global community, 47
    propositions, 19
    of social infrastructure, 47
Value orchestration platform model, 51
Ventures, 66
Viable systems approach, 77
Virtual, 66